P9-ARU-515

Lecture Notes in Mathematics

Edited by A. Dold and B. Eckmann

609

General Topology and Its Relations to Modern Analysis and Algebra IV

Proceedings of the Fourth Prague Topological Symposium, 1976

Part A: Invited Papers

Edited by J. Novák

Springer-Verlag
Berlin Heidelberg New York 1977

Editor
Josef Novák
Institute of Mathematics
of the ČSAV
Žitná 25
115 67 Praha 1
Czechoslovakia

Library of Congress Cataloging in Publication Data

Symposium on General Topology and Its Relations to
Modern Analysis and Algebra, 4th, Prague, 1976.
General topology and its relations to modern analysis
and algebra IV.

(Lecture notes in mathematics ; 609-)
Bibliography: p.
Includes index.
CONTENTS: pt. A. Invited papers.
1. Topology--Congresses. 2. Mathematical analysis--
Congresses. 3. Algebra--Congresses. I. Novák, Josef,
1905- II. Title. III. Series: Lecture notes
in mathematics (Berlin) ; 609.
QA3.L28 no. 609 [QA611] 510'.8s [514'.3] 77-11622

AMS Subject Classifications (1970): 54-06, 46-06, 28-06

ISBN 3-540-08437-1 Springer-Verlag Berlin Heidelberg New York
ISBN 0-387-08437-1 Springer-Verlag New York Heidelberg Berlin

Printed in Germany

Printing and binding: Beltz Offsetdruck, Hemsbach/Bergstr.
2141/3140-543210

PREFACE

Since 1961 every five years Prague topological symposium has been held by the Czechoslovak Academy of Sciences. In August 1976 the Fourth Symposium on General Topology and its Relations to Modern Analysis and Algebra took place in Prague. Arrangements for the symposium were made by the Organizing Committee consisting of J.Novák (chairman), Z.Frolík (program chairman), J.Hejcman, M.Hušek, M.Katětov, V.Koutník, V.Pták, A.Rázek (treasurer), M.Sekanina, Š.Schwarz, and V.Trnková. The Czechoslovak Academy of Sciences, the Slovak Academy of Sciences, the Faculty of Mathematics and Physics of the Charles University, and the Association of Czechoslovak Mathematicians and Physicists invited a number of prominent mathematicians from abroad. The International Mathematical Union granted financial support towards the travel expenses of some young mathematicians from abroad. I would like to thank the International Mathematical Union for all the support given to Prague topological symposia.

The program of the symposium consisted of 11 plenary invited talks, 18 invited talks given in two sections, 1 invited talk in a session for contributed papers and 134 fifteen-minute talks in three or four parallel sessions.

The symposium was attended by 217 mathematicians from 24 countries; 164 from abroad and 53 from Czechoslovakia.

These Proceedings of the symposium are published in two parts A and B. Part A, published by Springer-Verlag, contains 19 invited papers. Part B, published by the Association of Czechoslovak Mathematicians and Physicists, contains 105 abstracts and contributed papers.

I would like to thank the referees who read the manuscript. Special thanks are due to Z.Frolík, M.Katětov, V.Koutník and V.Pták for their assistance in editing these Proceedings.

J.Novák

LIST OF PARTICIPANTS

J.Albrycht (Poznań), A.Alexiewicz (Poznań), R.D.Anderson (Baton Rouge), A.V.Arhangel'skii (Moskva), S.P.Arya (New Delhi), M.Atsuji (Saitama), P.C.Baayen (Amsterdam), F.Bagi(Budapest), Z.Balogh (Debrecen), B.Banaschewski (Hamilton), V.Bartík (Praha), C.Behlivanidis (Athens), D.P.Bellamy (Newark), C.Bessaga (Warszawa), E.Binz (Mannheim), A.Błaszczyk (Katowice), J.Bočková (Praha), S.A.Bogatyj (Moskva), M.Bognár (Budapest), H.Boseck (Greifswald), H.Brandenburg (Berlin), L.Bukovský (Košice), D.W.Bushaw (Pullman), T.Byczkowski (Wrocław), Z.Čerin (Zagreb), J.Chaber (Warszawa), T.A.Chapman (Lexington), M.G.Charalambous (Zaria), J.J.Charatonik (Wrocław), G.Chogoshvili (Tbilisi), J.Chvalina (Brno), J.Činčura (Bratislava), W.W.Comfort (Middletown), Á.Császár (Budapest), K.Császár (Budapest), I.Dimitrov (Sofia), I.Dobrakov (Bratislava), D.Doitchinov (Sofia), L.Dokas (Patras), R.Domiaty (Graz), E.K.van Douwen (Athens), J.Dravecký (Bratislava), M.Duchoň (Bratislava), J.Durdil (Praha), R.Dyckhoff (Fife), Z.Dzedzej (Gdańsk), M.Fabian (Praha), M.Fiedler (Praha), H.R.Fischer (Amherst), J.Flachsmeyer (Greifswald), W.Fleissner (Montreal-Athens), R.Fox (London), R.Frič (Žilina), Z.Frolík (Praha), S.Gähler (Berlin), W.Gähler (Berlin), J.Gerlits (Budapest), G.Gierz (Darmstadt), W.Glowczyński (Gdańsk), W.Govaerts (Brussel), G.Grimeisen (Stuttgart), G.Gruenhage (Auburn), N.Hadjiivanov (Sofia), D.W.Hajek (Mayagüez), P.Hamburger (Budapest), K.Hardy (Ottawa), S.Heinrich (Berlin), J.Hejcman (Praha), N.Helderman (Berlin), M.Henriksen (Claremont), P.Hilton (Seattle), P.Holický (Praha), T.Hoshina (Ibaraki), M.Hušek (Praha), H.Husová (Praha), S.Iliadis (Patras), A.A.Ivanov (Leningrad), I.Ivanšić (Zagreb), A.Iwanik (Wrocław), J.E.Jayne (London), T.M.Jedryka (Bydgoszcz), K.John (Praha), B.E.Johnson (Newcastle), I.Juhász (Budapest), A.Kartsaklis (Athens), M.Katětov (Praha), F.Katrnoška (Praha), Y.Katuta (Ibaraki), S.Kaul (Regina), K.Keimel (Darmstadt), D.C.Kent (Pullman), G.Kneis (Berlin), P.A.Kolmus (Paris), J.Komorník (Bratislava), V.Kořínek (Praha), V.Koutník (Praha), J.Krasinkiewicz (Warszawa), P.Kratochvíl (Praha), P.B.Krikelis (Athens), P.Křivka (Pardubice), A.Kucia (Katowice), W.Kulpa (Katowice),K.Kuratowski (Warszawa), Đ.Kurepa (Beograd), V.Kůrková (Praha), V.I.Kuz'minov (Novosibirsk), O.V.Lokucievskij (Moskva), R.Lowen (Brussel), E.Lowen-Colebunders (Brussel), J.Lukeš (Praha), B.Lüschow (Berlin), T.Maćkowiak (Wrocław), G.de Marco (Padova), S.Mardešić (Zagreb), I.Marek (Praha), M.Marjanović (Beograd), A.Marquina (Valencia), G.Meletiou

(Arta), J.van Mill (Amsterdam), P.Minc (Warszawa), J.Mioduszewski (Katowice), L.Mišík (Bratislava), M.W.Mislove (New Orleans), J.Močkoř (Ostrava), B.Morrel (Amsterdam), B.Müller (Mannheim), K.Musiał (Wrocław), J.Musielak (Poznań), A.Mysior (Gdańsk), J.Nagata (Amsterdam), Z.Nagy (Budapest), S.A.Naimpally (Thunder Bay), S.J.Nedev (Sofia), S.Negrepontis (Athens), L.D.Nel (Ottawa), E.Nelson (Hamilton), J.Nešetřil (Praha), T.Neubrunn (Bratislava), J.Novák (Praha), P.J.Nyikos (Urbana), J.Pachl (Praha), C.M.Pareek (Kuwait), H.L.Patkowska (Warszawa), A.Patronis (Athens), J.Pechanec (Praha), J.Pelant (Praha), A.Pietsch (Jena), R.Pol (Warszawa), H.Poppe (Wustrow), H.Porst (Bremen), D.Preiss (Praha), G.Preuss (Berlin), T.Priftis (Peania), I.Prodanov (Sofia), P.Pták (Praha), V.Pták (Praha), A.Pultr (Praha), S.Purisch (Ibadan), H.Pust (Berlin), M.Rajagopalan (Memphis), B.el Rayess (Praha), G.M.Reed (Athens), H.C.Reichel (Wien), J.Reiterman (Praha), M.D.Rice (Fairfax), B.Riečan (Bratislava), J.R.Ringrose (Newcastle), J.Roitman (Wellesley), J.Rosický (Brno), M.E.Rudin (Madison), S.S.Ryškov (Moskva), E.V.Ščepin (Moskva), J.Schmets (Liège), A.Schrijver (Amsterdam), J.Schröder (Berlin), J.Segal (Seattle), M.Sekanina (Brno), M.Seyedin (Tehran), P.Simon (Praha), J.Šipoš (Bratislava), J.M.Smirnov (Moskva), M.Smith (Auburn), A.Šostak (Riga), P.Stavrinos (Athens), A.H.Stone (Rochester), D.Maharam Stone (Rochester), I.A.Švedov (Novosibirsk), B.Szőkefalvi-Nagy (Szeged), A.Szymański (Katowice), R.Talamo (Torino), F.D.Tall (Toronto), G.Tashjian (Hamilton), R.Telgársky (Wrocław), F.Terpe (Greifswald), W.J.Thron (Boulder), H.Toruńczyk (Warszawa), A.Triantafillou (Athens), V.Trnková (Praha), M.Turzański (Katowice), L.Vašák (Praha), J.E.Vaughan (Greensboro), M.L.van de Vel (Amsterdam), M.Venkataraman (Madurai), J.Vilímovský (Praha), R.Voreadou (Athens), J.de Vries (Amsterdam), M.Wage (New Haven - Madison), A.Waszak (Poznań), E.Wattel (Amsterdam), W.Weiss (Budapest), K.Wichterle (Praha), H.H.Wicke (Athens), M.Wilhelm (Wrocław), S.Willard (Edmonton), R.Wilson (Mexico), J.M.van Wouwe (Amsterdam), M.Zahradník (Praha), A.Zanardo (Padova), A.V.Zarelua (Tbilisi), D.Zaremba (Wrocław), J.Zemánek (Praha), P.Zenor (Auburn), S.P.Zervos (Athens), V.Zizler (Praha)

LIST OF INVITED ADDRESSES

Anderson,R.D.	Group actions on Hilbert cube manifolds
Arhangel´skiĭ,A.V.	Some recent results on cardinal-valued invariants of bicompact Hausdorff spaces
Bessaga,C.: Bessaga,C. and Dobrowolski,T.	Deleting formulas for topological vector spaces and groups (Presented by C.Bessaga)
Binz,E.	On an extension of Pontryagin´s duality Theory
Chapman,T.A.	Homotopy conditions which characterize simple homotopy equivalences
Comfort,W.W.	Some recent applications of ultrafilters to topology
Császár,Á.	Some problems concerning C(X)
Dobrowolski,T.: Bessaga,C. and Dobrowolski,T.	Deleting formulas for topological vector spaces and groups (Presented by C.Bessaga)
Van Douwen,E.K.	A technique for constructing examples
Efremovič,V.A.: Efremovič,V.A. and Vaĭnšteĭn,A.G.	Novye rezultaty v ravnomernoĭ topologii (Presented by O.V.Lokucievskij)
Flachsmeyer,J.	Topologization of Boolean algebras
Frolík,Z.	Recent development of theory of uniform spaces
Hilton,P.	On generalizations of shape theory
Hušek,M.: Hušek,M. and Trnková,V.	Categorial aspects are useful in topology
Johnson,B.E.	Perturbations of Banach algebras
Juhász,I.	On the number of open sets
Kuratowski,K.	σ-algebra generated by analytic sets and applications
Mardešić,S.	Recent development of shape theory
Nagata,J.	On rings of continuous functions

LIST OF COMMUNICATIONS

Patronis,A.: Stavrinos,P. and Patronis,A.	On hyperspace of continuous curves and continuous hyperspaces of any order (Presented by P.Stavrinos)
Pelant,J.: Pelant,J. and Reiterman,J.	Atoms in uniformities and proximities (Presented by J.Pelant)
Pol,E.: Pol,R. and Pol,E.	A hereditarily normal strongly zero--dimensional space containing subspaces of arbitrarily large dimension (Presented by R.Pol)
Pol,R.: Pol,R. and Pol,E.	A hereditarily normal strongly zero--dimensional space containing subspaces of arbitrarily large dimension (Presented by R.Pol)
Poppe,H.: Jeschek,F., Poppe,H. and Stärk,A.	A compactness criterion for the space of almost periodic functions (Presented by H.Poppe)
Porst,H.E.	Embeddable spaces and duality
Preuss,G.	Relative connectedness and disconnectedness in topological categories
Prodanov,I.	Minimal topologies on Abelian groups
Pták,P.	Refinements of uniform spaces associated with discreteness
Reed,G.M.	Consistency and independence results in Moore spaces
Reichel,H.C.	Some results on distance-functions
Reiterman,J.: Pelant,J. and Reiterman,J.	Atoms in uniformities and proximities (Presented by J.Pelant)
Rice,M.D.	Descriptive sets in uniform spaces
Riečan,B.	Extension of measures and integrals by the help of a pseudometric
Richardson,G.D.: Kent,D.C. and Richardson,G.D.	Some product theorems for convergence spaces (Presented by D.C.Kent)

Terpe,F.: Flachsmeyer,J. and Terpe,F.	On convergence structures in the space of summations (Presented by F.Terpe)
Thron,W.J.: Chattopadhyay,K.C. and Thron,W.J.	Extensions of closure spaces (Presented by W.J.Thron)
Vaughan,J.E.	Products of $[a,b]$ -chain compact spaces
Van de Vel,M.	The fixed point property of superextensions
Venkataraman,M.	On topological invariants of topological spaces
Vilímovský,J.	Extensions of uniformly continuous Banach valued mappings
De Vries,J.	Embeddings of G-spaces
Wage,M.L.	On a problem of Katětov
Waszak,A.: Musielak,J. and Waszak,A.	Some problems of convergence in countably modulared spaces (Presented by J.Musielak)
Wattel,E.	Subbase structures in nearness spaces
Weiss,W.	Some connected problems
Wichterle,K.	Joint convergence in function spaces. Order of W-closures
Wicke,H.H.: Wicke,H.H. and Worell,J.M.,Jr.	Scattered spaces of point-countable type (Presented by H.H.Wicke)
Wilhelm,M.	On closed graph theorems
Worell,J.M.,Jr.: Wicke,H.H. and Worell,J.M.,Jr.	Scattered spaces of point-countable type (Presented by H.H.Wicke)
Van Wouwe,J.M.	Generalized ordered p-spaces
Zahradník,M.	ℓ_1-partitions of unity on normed spaces
Zarelua,A.	On zero-dimensional mappings and commutative algebra
Zaremba,D.	On cohesive mappings
Zemánek,J.	On the spectral radius in Banach algebras
Zenor,P.: Alster,K. and Zenor,P.	Preservation of the Lindelöf property in product spaces (Presented by P.Zenor)

CONTENTS

On an Extension of Pontryagin's Duality Theory

E.Binz
Mannheim

1. Introduction

Let G be a commutative, topological group. A character of G is a continuous homomorphism $h : G \longrightarrow S^1$, where the group S^1 is the compact group of all complex numbers of modulus one. Now let G be locally compact. The collection ΓG of all characters of G, endowed with the topology of compact convergence, forms a commutative, locally compact, topological group $\Gamma_c G$ under the pointwise defined operations. In addition, the natural homomorphism

$$j_G : G \longrightarrow \Gamma_c \Gamma_c G ,$$

defined by $j_G(g)(\gamma) = \gamma(g)$ for each $g \in G$ and each $\gamma \in \Gamma_c G$, is, as the fundamental theorem of Pontryagin states, a bicontinuous isomorphism. Pontryagin's duality theory is the study of the rich relations between G and $\Gamma_c G$.

The aim of this note is to suggest an extension of Pontryagin's duality theory by extending the fundamental theorem to a wider class of groups. We proceed as follows: To the (commutative) groups under consideration will be associated a concept of convergence compatible with the algebraic structure. Groups of this sort are called convergence groups. This concept of convergence, given by a convergence structure, will allow the notion of continuity. For any convergence group G, the group ΓG of all characters of G equipped with the continuous convergence structure Λ_c will be denoted by $\Gamma_c G$. In case G is a locally compact topological group, Λ_c is identical to the topology of compact convergence. We will call a convergence group

P_c-reflexive, if $j_G : G \longrightarrow \Gamma_c\Gamma_c G$ is a bicontinuous isomorphism. The class of P_c-reflexive convergence groups contains in addition to all commutative, locally compact, topological groups all the complete, locally convex $\mathbb{R}$-vector spaces. We will verify the P_c-reflexivity of the following type of topological groups:

For any $k = 0,\ldots,\infty$ the collection $C^k(M,S^1)$ of all S^1-valued C^k-functions of a connected compact C^∞-manifold M, equipped with the C^k-topology is a topological group. It is in general not locally compact. We demonstrate the P_c-reflexivity of $C^k(M,S^1)$ as follows:

The idea is to use $C^k(M)$, the complete, locally convex vector space of real-valued C^k-functions of M equipped with the C^k-topology and to introduce $C^k(M)/\underline{\mathbb{Z}}$, where $\underline{\mathbb{Z}}$ denotes the subset of all the functions assuming their values in $\mathbb{Z}$. We will show, that $C^k(M)/\underline{\mathbb{Z}}$ can be identified with the connected component $\varkappa_M C^k(M)$ of $\underline{1}$ in $C^k(M,S^1)$. The quotient $C^k(M,S^1)/\varkappa_M C^k(M)$ is then a discrete group called $\Pi^1(M)$. The exact sequence

$$\underline{1} \longrightarrow \varkappa_M C^k(M) \longrightarrow C^k(M,S^1) \longrightarrow \Pi^1(M) \longrightarrow 1$$

has an exact "bidual":

$$\underline{1} \longrightarrow \Gamma_c\Gamma_c\varkappa_M C^k(M) \longrightarrow \Gamma_c\Gamma_c C^k(M,S^1) \longrightarrow \Gamma_c\Gamma_c\Pi^1(M) \longrightarrow \underline{1}.$$

Since $\varkappa_M C^k(M)$ and $\Pi^1(M)$ will turn out to be P_c-reflexive, we will conclude, via the five lemma, that $C^k(M,S^1)$ is also P_c-reflexive. Along the way, we will study some special character groups appearing in our procedure.

For this type of extension of Pontryagin's duality theory a suitable extension theorem of characters is still missing. This hinders considerably the study of the relations between G and $\Gamma_c G$ for P_c-reflexive convergence groups G .

2. Review of some Definitions and Results

2.1. The character group of a convergence group, P_c-reflexivity

Let X be a non empty set. To any point in X will be associated a collection $\Lambda(p)$ of filters on X. The set $\Lambda(p)$ is an element of $\mathcal{P}(F(X))$, the power set of the set of all filters $F(X)$ of X.

The map $\Lambda : X \longrightarrow \mathcal{P}(F(X))$ is called a convergence structure on X if the following conditions are satisfied for each $p \in X$:

(i) $\dot{p}$, the filter generated by $\{p\}$ belongs to $\Lambda(p)$.

(ii) Any filter finer than a member of $\Lambda(p)$ belongs to $\Lambda(p)$.

(iii) The infimum $\Phi \wedge \psi$ of any two filters of $\Lambda(p)$ belongs to $\Lambda(p)$.

Let us remark here, that any topology on X is a convergence structure, but not vice versa.

The set X, together with a convergence structure Λ, is called a convergence space. The filters in $\Lambda(p)$ are said to converge to p in X. A map f from a convergence space X into a convergence space Y is continuous if, for any filter Φ convergent to p in X, the image converges to $f(p)$ in Y. The cartesian product $X \times X$ of any two convergence spaces X and Y carries the product structure defined in the obvious way [Bi].

On $C(X,Y)$, the collection of the continuous maps from the convergence space X into the convergence space Y, there is a coarsest among all the convergence structures for which the evaluation map

$$\omega : C(X,Y) \times X \longrightarrow Y ,$$

(defined by $\omega(f,p) = f(p)$ for any $(f,p) \in C(X,Y) \times X$) is continuous. This is called the continuous convergence structure Λ_c. A filter Θ on $C(X,Y)$ converges to a function f with respect to Λ_c iff for any $p \in X$ the filter $\omega(\Theta \times \Phi)$ converges to $f(p)$

in Y for any filter Φ convergent to p. The set $C(X,Y)$ and any subset $A(X,Y)$ of $C(X,Y)$ endowed with Λ_c are denoted by $C_c(X,Y)$ and $A_c(X,Y)$ respectively. The continuous convergence structure is characterized by the following universal property ([Bi], [Bi,Ke]) : A map f from a convergence space S into a subspace $A_c(X,Y)$ of $C_c(X,Y)$ is continuous iff $\omega \circ (f\times id): S \times X \longrightarrow Y$ is continuous.

We now pass on to convergence groups. Our groups are always assumed to be abelian.

A group, together with a convergence structure, is called a <u>convergence group</u> if the group operations are continuous.

The character group $\Gamma_c G$ of a convergence group G is the group ΓG of all continuous homomorphisms of G into the circle group S^1 together with the continuous convergence structure. The operations on ΓG are defined pointwise. Obviously, $\Gamma_c G$ is a convergence group. The canonical map

$$j_G : G \longrightarrow \Gamma_c\Gamma_c G ,$$

defined by $j_G(g)(\gamma) = \gamma(g)$ for any $g \in G$ and any $\gamma \in \Gamma_c G$ is evidently continuous.

We call G P_c-<u>reflexive</u> if j_G is a bicontinuous isomorphism

<u>Remark:</u> If G is a locally compact topological group, then the continuous convergence structure on ΓG is identical to the topology of compact convergence. Hence the P_c-reflexivity of such a group G is identical to the classical reflexivity in the sense of Pontryagin [Po].

<u>2.2 The character group of a convergence vector space</u>

An $\mathbb{R}$-vector space E (referred to as a vector space) equipped with a convergence structure for which the operations are continuous is called a <u>convergence vector space</u> [Bi]. The <u>c-dual</u>, $\mathcal{L}_c E$, of E is the vector space of all continuous real-valued linear functionals endowed with the continuous convergence structure.

The exponential map from $\mathbb{R}$ to S^1 sending each real r to $e^{2\pi ir}$ is denoted by $\varkappa$. This map induces a continuous homomorphism $\varkappa_* : L_cE \longrightarrow \Gamma_cE$ assigning to each $\ell \in L_cE$ the character $\varkappa \circ \ell$. It is shown in [Bu], that $\varkappa_*$ is a bicontinuous isomorphism. For a slightly restricted version of this result, which is general enough for our purposes, we refer the reader to the Appendix in [Bi]. The proof of the result in [Bu] is an elaborated version of the proof I gave in [Bi]. For an earlier result in this direction see [F-S].

Let us point out here, that there is no vector space topology T on LE, where E is locally convex, for which the evaluation map $\omega : LE \times E \longrightarrow \mathbb{R}$ is continuous, unless E is normable [Ke].

We call a convergence vector space E c-reflexive if $i_E : E \longrightarrow L_cL_cE$ is a bicontinuous isomorphism.

One easily verifies [Bi]:

Lemma 1: A convergence vector space E is P_c-reflexive iff E is c-reflexive. A topological vector space E is P_c-reflexive iff it is locally convex and complete.

2.3. P_c-reflexivity of some convergence groups of continuous mappings

Assume that X is an arcwise connected topological space. The map $\varkappa_X : C_c(X) \longrightarrow C_c(X,S^1)$, sending each $f \in C_c(X)$ into $\varkappa \circ f$, is a quotient map onto its range, regarded as a subspace of $C_c(X,S^1)$ [Bi,2]. The quotient $C_c(X,S^1)/\varkappa_X C_c(X)$, carrying the quotient structure is denoted by $\Pi_c^1(X)$ and is called [Hu] the Bruschlinski group of X. If X is locally compact, $C_c(X,S^1)$ is a topological group. As demonstrated in [Bi,2], we have:

Theorem 2 The group $\varkappa_X C_c(X)$ is P_c-reflexive. If, in addition, X is a normal space allowing a (simply connected) universal covering, the group $C_c(X,S^1)$ is P_c-reflexive if $\Pi_c^1(X)$ is complete. This is the case e.g. if either the first singular homology group (with the integers as coefficients) is finitely generated, or $\Pi^1(X)$ is isomorphic to the first singular cohomology group (with the integers as coefficients).

In the next few sections we will derive some functional analytic results which will, in turn, be fundamental in showing the P_c-reflexivity of $C^k(M,S^1)$, as announced in the introduction.

3. Functional analytic preliminaries

3.1. $C^k(M)$ for a connected compact C^∞-manifold M

Let M be a compact C^∞-manifold. For a non-negative integer k, we will denote by $C^k(M)$ the Banach space of all real-valued C^k-functions of M, endowed with the usual norm. This yields the topology of uniform convergence in all k derivatives. We refer to [Pa] and [Go,Gui] for the above remarks and for the next few details. Clearly the inclusion map $j_k^{k+1} : C^{k+1}(M) \longrightarrow C^k(M)$ is continuous for any k. Moreover, its image is dense and the image of the unit ball E_{k+1} of $C^{k+1}(M)$ is relatively compact in $C^k(M)$.

The projective limit of all $C^k(M)$ is denoted by $C^\infty(M)$. This is a complete, metrizable, locally convex space, a so-called Fréchet space [Schae]. Since E_{k+1} is relatively compact in $C^k(M)$ for any k, the space $C^\infty(M)$ is called a Schwartz space.

3.2. The c-dual of $C^k(M)$

First, let F be any convergence vector space. Any compact set in $\mathcal{L}_c F$ is topological [Bi]. A convergence space is said to be compact if every ultrafilter converges to exactly one point.

Next, we describe the c-dual of F where F is a topological vector space. For any neighborhood U of zero in F, the polar $\{\ell \in LF \mid \ell(U) \subset [-1,1]\}$, denoted by U^o, is compact if regarded as a subspace of L_cF. Hence it is a compact topological space. The topology on it is the topology of pointwise convergence. Moreover, L_cF is the inductive limit (in the category of convergence spaces) of all these compact topological spaces U^o, where U runs through the neighborhood filter of zero in F. For these and the next few details we refer the reader to [Bi] or to [Bi,Bu,Ku].

For a topological vector space F, the natural map $i_F : F \longrightarrow L_cL_cF$ is a bicontinuous isomorphism iff F is a complete, locally convex vector space (cf. Lemma 1).

Let us turn our attention to $L_cC^k(M)$ for a finite k. The convergence vector space $L_cC^k(M)$ is the inductive limit of all multiples of the polar E_k^o of the unit ball $E_k \subset C^k(M)$. Here as a subspace of $L_cC^k(M)$, E_k^o carries the topology of pointwise convergence and is therefore compact.

When $LC^k(M)$ carries the usual norm topology, we write $L_nC^k(M)$.

For any k, the adjoint of j_k^∞, the map

$$j_k^{\infty *} : L_cC^k(M) \longrightarrow L_cC^\infty(M) ,$$

defined by composing each $\ell \in L_cC^k(M)$ with j_k^∞, is a continuous injection. Since $C^\infty(M)$ is a Schwartz space, we even have [Ja] :

<u>Lemma 3</u> $L_cC^\infty(M)$ is the inductive limit (in the category of convergence spaces) of $L_cC^k(M)$ as well as of $L_nC^k(M)$, taken over all finite k.

For any $p \in M$, the linear functional $i_M(p) : C^k(M) \longrightarrow \mathbb{R}$ evaluating each $f \in C^k(M)$ at p is continuous for any k. If $k < \infty$ $i_M^k : M \longrightarrow L_cC^k(M)$ sending each $p \in M$ into $i_M^k(p)$ is a continuous injection whose image is contained in the polar E_k^o of the unit ball $E_k \subset C^k(M)$. Hence we have:

Lemma 4: The canonical map $i^k_M : M \longrightarrow L_c C^k(M)$ is (for any k) a homeomorphism onto a subspace of $L_c C^k(M)$. If $k < \infty$, then $i^k_M(M) \subset E^o_k$.

3.3. $V^k(M)$

For each $k=0,1,\ldots,\infty$ let $V^k(M)$ be the span of $i^k_M(M)$, regarded as a subspace of $L_c C^k(M)$.

Recall that in a convergence space X a point p is adherent to a subset A if there is a filter Φ convergent to p in X, such that $F \cap A \neq \emptyset$ for any $F \in \Phi$. We call $A \subset X$ dense if the collection $\bar{A}$ (the adherence of A) of all points adherent to A is all of X.

The following is an analogue to the situation in the case of $C_c(X)$ (cf. appendix of [Bi]).

Theorem 5: The space $V^\infty(M)$ is dense in $L_c C^\infty(M)$. Moreover the restriction map $r^\infty : L_c L_c C^\infty(M) \longrightarrow L_c V^\infty(M)$ is a bicontinuous isomorphism. Thus $\rho^\infty : L_c V^\infty(M) \longrightarrow C^\infty(M)$, defined by $\rho^\infty(\ell) = \ell \circ i^\infty_M$ for each $\ell \in L_c V^\infty(M)$, is a bicontinuous isomorphism.

Proof: Since $C^\infty(M)$ is c-reflexive, r^∞ is a monomorphism. To show its surjectivity, consider for each finite k the following diagram:

$$V^\infty(M) \xleftarrow{\ a\ } V^{k+1}_n(M) \subset L_n C^{k+1}(M) \xleftarrow{(j^{k+1}_k)^*} L_c C^k(M) .$$

The index n indicates, that the respective spaces carry the usual norm topology. The linear map a sending each $i^{k+1}_M(p)$ into $i^\infty_M(p)$, is evidently continuous. Finally $(j^{k+1}_k)^*$ restricts each $\ell \in L_c C^k(M)$ to $C^{k+1}(M)$. The next goal is to show that $(j^{k+1}_k)^*$ is continuous. We recall that the unit ball E_{k+1} of $C^{k+1}(M)$ is relatively compact in $C^k(M)$. Hence the polar E^o_k of E_k formed in $LC^k(M)$ is mapped by $(j_k^{k+1})^*$ into a compact subspace $(j_k^{k+1})^*(E_k^o)_n$ of the Banach space $L_n C^{k+1}(M)$ (cf. [Schae] p.111). From this, we conclude the continuity of $(j_k^{k+1})^*$. Next let $\ell \in L_c V^\infty(M)$. The functional $\ell \circ a$ has a continuous extension $\bar{\ell}$ to $L_n C^{k+1}(M)$, for which $\bar{\ell} \circ (j_k^{k+1})^* = \ell'$ is continuous on $L_c C^k(M)$. Moreover $\ell' \circ i_M^k = \ell \circ i_M^\infty$. Since $C^k(M)$ is c-reflexive, ℓ' can be represented as $i_{C^k(M)}(f_k)$ for some function $f_k \in C^k(M)$. Hence $\ell' \circ i_M^k = f_k$ for each finite k. Since $\ell \circ i_M^\infty = f_k$, the function $\ell \circ i_M^\infty$ is of class C^∞ and $r^\infty \circ i_{C^\infty(M)}(f_k) = \ell$. Thus the injection is bijective. We proceed now to show that $V^\infty(M)$ is dense in $L_c C^\infty(M)$. To do this, we introduce $\overline{V^\infty(M)}'$ and establish three properties (a,b,c) which exhibit this space as an L_c-space [Bi,Bu,Ku]. Thus $\overline{V^\infty(M)}'$ is c-reflexive. Let us point out that $L_c C^\infty(M)$ is the inductive limit (in the category of convergence spaces) of countably many absolutely convex, compact topological spaces $K_1 \subset K_2 \subset \dots$. Hence we have $\bigcup_i K_i = L C^\infty(M)$. For each index i we form the adherence $\overline{K_i \cap V^\infty(M)}^i$ of $K_i \cap V^\infty(M)$ in K_i. This adherence is a convex, compact topological subspace of K_i. Moreover

$$\overline{V^\infty(M)} = \bigcup_i \overline{K_i \cap V^\infty(M)}^i ,$$

as one easily verifies. Hence $\overline{V^\infty(M)}'$, regarded as the inductive limit of the compact, convex subspaces $\overline{K_i \cap V^\infty(M)}^i$, taken over all i, is a) locally convex, and locally compact and b) admits point-separating continuous linear functionals. By locally convex we mean that, for any filter convergent to q, there is a coarser one having a basis of convex sets which also converges to q. Locally compact means that any convergent filter contains a compact set.

The last one, c), of the above mentioned characteristic properties is the following: Any compact subspace of $\overline{V^\infty(M)}'$ is a compact topological space. But this is evidently true because any compact subset of $\overline{V^\infty(M)}'$ is contained in one of the compact topological spaces

$\overline{K_i \cap V^\infty(M)^i}$. Thus $\overline{V^\infty(M)}'$ is an L_c-space. Since $\overline{V^\infty(M)}'$ splits into countably many compact subsets, $L_c\overline{V^\infty(M)}'$ is a Fréchet space. Hence it is c-reflexive [Bi,Bu,Ku]. In addition, $V^\infty(M)$ is a dense subspace of $\overline{V^\infty(M)}'$. One easily concludes that

$$r^\infty : L_cL_cC^\infty(M) \longrightarrow L_c\overline{V^\infty(M)}'$$

is a continuous bijection between Fréchetspaces. Using the closed graph theorem, we deduce that r^∞ is a homeomorphism. The c-reflexivity of $\overline{V^\infty(M)}'$ and $C^\infty(M)$ now immediately yields $\overline{V^\infty(M)} = L_cC^\infty(M)$. The commutativity of

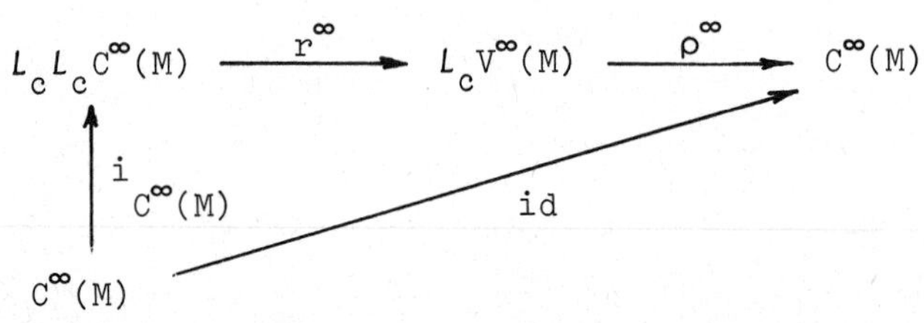

allows us to conclude that r^∞ and ρ^∞ are bicontinuous isomorphisms, as asserted in theorem 5.

4. $\varkappa_M C^k(M)$, in particular $\varkappa_M C^\infty(M)$

4.1 The group $\varkappa_M C^k(M)$ and its P_c-reflexivity

For any $k = 0,\ldots,\infty$, we consider the collection $\varkappa_M C^k(M)$ of all functions $\varkappa \circ f$, where $f \in C^k(M)$. (Recall, that $\varkappa : \mathbb{R} \longrightarrow S^1$ sends each r into $e^{2\pi i r}$). This collection is a group under the pointwise defined operations. Since M is connected, the kernel of

$$\varkappa_M : C^k(M) \longrightarrow \varkappa_M C^k(M)$$

is $\underline{\mathbb{Z}}$, the subgroup of all constant functions assuming their values in $\mathbb{Z}$. For any $z \in \mathbb{Z}$ denote by $\underline{z}$ the function whose only value is z . By virtue of the addition in $C^k(M)$, $\mathbb{Z}$ operates on $C^k(M)$ properly discontinuously [Spa]. Hence the quotient $C^k(M)/\underline{\mathbb{Z}}$, taken in the category of topological spaces, has $C^k(M)$ as its (simply connected) universal covering. From this, we conclude that $C^k(M)/\mathbb{Z}$ is also the quotient in the category of convergence spaces. Moreover $\mathbb{Z}$

is isomorphic to the fundamental group of $C^k(M)/\mathbb{Z}$. Let us identify $C^k(M)/\mathbb{Z}$ with $\varkappa_M C^k(M)$ and the projection map onto $C^k(M)/\mathbb{Z}$ with $\varkappa_M$.

The topological group $\varkappa_M C^k(M)$ can be represented as a direct product (I thank H.P.Butzmann for reminding me of this fact): To a given point $p \in M$ consider the subspace $m_p^k \subset C^k(M)$ consisting of all C^k-functions vanishing on p. For any two functions $f_1, f_2 \in m_p^k$ we have $f_1 - f_2 \notin \mathbb{Z}$ unless they are identical. Hence $\varkappa_M | m_p^k$ is an injection and we conclude from the topological direct sum decomposition $C^k(M) = m_p^k \oplus \mathbb{R} \cdot \underline{1}$ that

$$\varkappa_M C^k(M) = \varkappa_M(m_p^k) \cdot S^1 \cdot \underline{1} .$$

This direct decomposition is evidently topological. Since $m_p \subset C^k(M)$ is a complete, locally convex topological vector space, it is P_c-reflexive (Lemma 1). Since S^1 is also P_c-reflexive, $\varkappa_M C^k(M)$ is P_c-reflexive. Thus we have:

<u>Theorem 6</u> For any $k = o, \ldots, \infty$, the topological group $\varkappa_M C^k(M)$ has $C^k(M)$ as its universal covering with a fundamental group isomorphic to $\mathbb{Z}$, splits topologically into

$$\varkappa_M C^k(M) = \varkappa_M(m_p^k) \cdot S^1 \cdot \underline{1} .$$

and is thus P_c-reflexive.

4.2. The character group of $\varkappa_M C^\infty(M)$; the group $P_c^\infty(M)$

A linear combination $\Sigma r_i \cdot i_M^\infty(p_i) \in V^\infty(M)$ composed with $\varkappa$ factors through $\varkappa_M$ iff $\Sigma r_i \in \mathbb{Z}$. Denote by $P_c^\infty(M)^1 \subset \Gamma_c C^\infty(M)$ the collection of all combinations of the form $\varkappa \circ \Sigma r_i \cdot i_M^\infty(p_i)$ for which $\Sigma r_i \in \mathbb{Z}$, equipped with the continuous convergence structure. Since $\varkappa_M$ is a quotient map, the continuous homomorphism

$$\bar{\varkappa} : P_c^\infty(M)^1 \longrightarrow \Gamma_c \varkappa_M C^\infty(M) ,$$

assigning to each character in $P_c^\infty(M)^1$ its factorization through $\varkappa_M$, is a bicontinuous isomorphism onto a convergence subgroup of $\Gamma_c \varkappa_M C^\infty(M)$.

Denoting this convergence subgroup by $P_c^\infty(M)$, then we have a bicontinuous isomorphism

$$\bar{\varkappa} : P_c^\infty(M)^1 \longrightarrow P_c^\infty(M) .$$

Lemma 7 $P_c^\infty(M)$ is dense in $\Gamma_c \varkappa_M C^\infty(M)$.

The proof is analogous to that of Lemma 8 (p.67) given in [Bi,2] .

We may reformulate Lemma 7 by saying that the character group $\Gamma_c \varkappa_M C^k(M)$ of $\varkappa_M C^k(M)$ is generated by $P_c^\infty(M)$.

Next consider the injective mapping

$$j_M^\infty : M \longrightarrow P_c^\infty(M) \subset \Gamma_c \varkappa_M C^\infty(M)$$

defined by $j_M^\infty(p)(t) = t(p)$ for all $p \in M$ and all $t \in \varkappa_M C^\infty(M)$. Since $\varkappa_M$ is a quotient map, we conclude by Lemma 4, that j_M^∞ maps M homeomorphically onto a subspace of $P_c^\infty(M)$. Any character $\gamma \in \Gamma P_c^\infty(M)$ induces an S^1-valued function $\gamma \circ j_M^\infty$.

Lemma 8 For each $\gamma \in \Gamma P_c^\infty(M)$ the function $\gamma \circ j_M^\infty$ belongs to $\varkappa_M C^\infty(M)$
The map

$$\varphi : \Gamma_c P^\infty(M) \longrightarrow \varkappa_M C^\infty(M)$$

sending each γ into $\gamma \circ j_M^\infty$ is a continuous monomorphism.

Proof: For $\gamma \in \Gamma P_c^\infty(M)$ consider $\gamma \circ \bar{\varkappa} \in \Gamma P_c^\infty(M)^1$; and denote $\varkappa_*^{-1}(P_c^\infty(M)^1)$ by $V_1^\infty(M) \subset L_c C^\infty(M)$. Pulling the character $\gamma \circ \bar{\varkappa}$ back onto $V_1^\infty(M)$, we obtain the character $\gamma \circ \bar{\varkappa} \circ (\varkappa_* | V_1^\infty(M)) : V_1^\infty(M) \longrightarrow S^1$. Our aim is to extend this character onto the whole space $V^\infty(M)$ and then (using theorem 5) to show that $\gamma \circ j_M^\infty$ is of class C^∞. For this purpose we decompose $V^\infty(M)$ as follows: One factor is M_o , the kernel of the linear functional $i_{C^\infty(M)}(\underline{1}) : V^\infty(M) \longrightarrow \mathbb{R}$, sending each linear combination $\Sigma r_i \cdot i_M^\infty(p_i)$ into Σr_i. Hence M_o consists of all linear combinations $\Sigma_i r_i \cdot i_M^\infty(p_i)$ with $\Sigma r_i = o$. For a fixed point $p \in M$, we form $\mathbb{R}\cdot i_M^\infty(p)$, which is homeomorphic to $\mathbb{R}$. One easily shows now that

$$\text{(i)} \qquad V^\infty(M) = M_o \oplus \mathbb{R}\cdot i_M^\infty(p)$$

holds as an identity between convergence vector spaces. Hence $V_1^\infty(M) \subset V^\infty(M)$ decomposes as

$$\text{(ii)} \qquad V_1^\infty(M) = M_o \oplus \mathbb{Z}\cdot i_M^\infty(p) .$$

(An analoguous decomposition holds for any k.) We therefore split $\gamma \circ \bar{\varkappa} \circ (\varkappa_* | V_1^\infty(M))$ into the product $\gamma_1 \cdot \gamma_2$ of its restrictions $\gamma_1 = \gamma \circ \bar{\varkappa} \circ (\varkappa_* | M_o)$ and $\gamma_2 = \gamma \circ \bar{\varkappa} \circ (\varkappa_* | \mathbb{Z}\cdot i_M^\infty(p))$. Using the classical extension theorem of characters, we extend $\gamma_2 : \mathbb{Z}\cdot i_M^\infty(p) \longrightarrow S^1$ to $\bar{\gamma}_2 : \mathbb{R}\cdot i_M^\infty(p) \longrightarrow S^1$. Then $\gamma_1 \cdot \bar{\gamma}_2$ is a continuous character on $V^\infty(M)$ which corresponds via $\varkappa_*$ to a continuous linear functional $\ell \in L_c V^\infty(M)$. By theorem 5, the functional ℓ is of the form $(\rho^\infty)^{-1}(f)$ where $f \in C^\infty(M)$. From this we conclude $\varphi(\gamma) = \varkappa_M(f)$. Since ℓ is uniquely determined by its values on $i_M^\infty(M) \subset V^\infty(M)$, the continuous map φ is a monomorphism. This completes the proof. (The methods used above yield simplifications in the proof of Satz 7, p.62 in [Bi,2].)

Finally, let us collect some of our results on $\varkappa_M C^\infty(M)$ and its character group in the following theorem.

<u>Theorem 9</u> The topological group $\varkappa_M C^k(M)$ splits topologically into $\varkappa_M(m_p^k)\cdot S^1 \cdot \underline{1}$ where $m_p^k \subset C^k(M)$ consists of all C^k-functions vanishing on a fixed point $p \in M$. The character group of $\varkappa_M C^\infty(M)$ is generated by $P_c^\infty(M)$. Moreover $\varphi : \Gamma_c P_c^\infty(M) \longrightarrow \varkappa_M C^\infty(M)$, sending each character γ into $\gamma \circ j_M^\infty$, is a bicontinuous isomorphism. In addition, $P_c^\infty(M)$ splits topologically into $\bar{\varkappa}(N_o)\cdot \mathbb{Z}\cdot j_M(p)$, where N_o carries the continuous convergence structure and consists of all combinations $\varkappa \circ \Sigma r_i \cdot i_M^\infty(p_i) \in \Gamma C^\infty(M)$ with $\Sigma r_i = o$. The character group of $\bar{\varkappa}(N_o)$ is bicontinuously isomorphic to $\varkappa_M(m_p^\infty)$.

<u>Proof:</u> The first two assertions are valid by theorem 1 and Lemma 7. To verify the others, consider the commutative diagram of continuous maps:

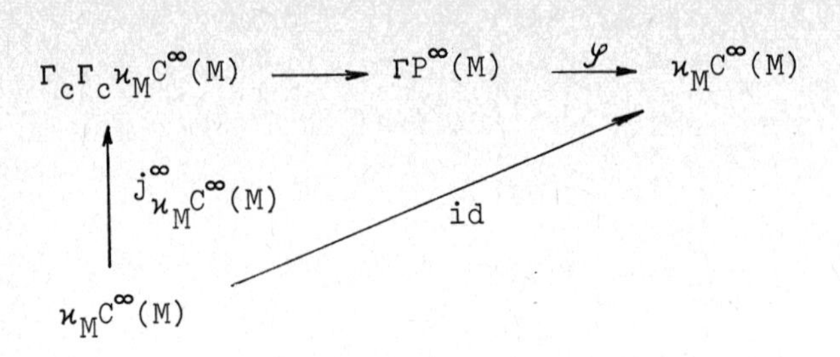

where the first horizontal arrow indicates the restriction map. Using this diagram in combination with Lemmas 7 and 8, we easily obtain the bijectivity of φ, the continuity of φ^{-1} and thus the P_c-reflexivity of $\varkappa_M C^\infty(M)$ again. That $P_c^\infty(M)$ splits into $\bar{\varkappa}(N_o)\cdot\mathbb{Z}\cdot j_M^k(p)$ is evident by using (ii) in the proof above and $\bar{\varkappa}$ introduced at the beginning of section 4.2. The rest of the theorem is straightforward.

5. $C^k(M,S^1)$ and its P_c-reflexivity

5.1. The Bruschlinski group

The collection of all S^1-valued C^k-functions endowed with the C^k- topology [Go,Gui] forms a topological group under the pointwise defined operations. For $k = o$, the topological group $C^o(M,S^1)$ carries the topology of compact convergence. In addition, $C^k(M,S^1)$ is a Banach manifold for each finite k and is a Fréchet manifold for $k = \infty$. (cf. [Go,Gui] p.76) . However, let us describe a canonical chart of the unit element $\underline{1}$. Consider in $S^1 \times \mathbb{R}$ (the tangent bundle of S^1), the neighborhood $S^1 \times (-1,1)$ of $S^1\times\{o\}$. The set $\Phi(U)$ of all functions $f \in C^k(M)$ for which $(\underline{1},f)(M) \subset S^1\times(-1,1)$ forms an <u>open</u> <u>convex</u> subset of $C^k(M)$. On the other hand, the set U defined by $\{\varkappa\circ f \mid f \in \Phi(U)\}$ is open in $C^k(M,S^1)$ and $\Phi^{-1}:\Phi(U) \longrightarrow U$ assigning to each $f \in \Phi(U)$ the map $\varkappa\circ f \in U$ is a homeomorphism. Observe, now, that $\Phi(U)$ is a subspace of $\varkappa_M C^k(M)$. In fact $\Phi(U) \subset \varkappa_M C^k(M)$ is evenly covered in $C^k(M)$, the universal covering of $\varkappa_M C^k(M)$ as remarked in § 4.1. Since $\Phi(U)$ is a subspace of both $C^k(M,S^1)$ and $\varkappa_M C^k(M)$, the topological group $\varkappa_M C^k(M)$ is an open topological subgroup of $C^k(M,S^1)$. It is even the connected component W of $\underline{1} \in C^k(M,S^1)$. This can be seen as follows. As a manifold, modeled on convex charts, W is pathwise connected. Any path $\sigma:[-1,1] \to W$ in W starting at $\underline{1}$ and ending at t defines a homotopy $\sigma:[-1,1] \times M \to S^1$, connecting $\underline{1}$ and t. Without loss of generality we may assume that $t(p_o)=1$. Thus both $\underline{1}$ and $\underline{t}$ define the trivial homomorphism from the fundamental group $\Pi_1(M,p_o)$ of M into $\Pi_1(S,1)$ the fundamental group of S^1. But this means that $t \in \varkappa_M C^k(M)$. (In case of $C^o(M,S^1)$, we used the compactness of M only but not the differentiable structure; no arcwise connectedness is needed either, compare § 2.3)

Let us form the quotient $C^k(M,S^1)/\varkappa_M C^k(M)$, whose quotient structure is the discrete topology.

We have the following

Lemma 10 For any $k = o,1\ldots,\infty$, the group $C^k(M,S^1)$ is dense in $C^o(M,S^1)$. Hence the inclusion $C^k(M,S^1) \subset C^o(M,S^1)$ induces an isomorphism $B:C^k(M,S^1)/\varkappa_M C^k(M) \longrightarrow C^o(M,S^1)/\varkappa_M C^o(M)$.

Proof: Consider $t \in C^o(M,S^1)$ and a map $(t,f) : M \to S^1 \times \mathbb{R}$ in the canonical chart of t, where $f \in C^o(M)$ composed with $\varkappa:\mathbb{R} \to S^1$ yields t. The set $C^\infty(M)$ is dense in $C^o(M)$. Hence we find a map $f_t \in C^\infty(M)$ close to f. But then $\varkappa \circ f_t$ is close to t, which proves the first assertion of the lemma. The second is a simple consequence.

As mentioned in § 2.3, we denote the quotient $C^o(M,S^1)/\varkappa_M C^o(M)$ by $\Pi^1(M)$. For any k, consider the homomorphism $b:C^k(M,S^1) \to \Pi^1(M)$, which is the canonical projection onto $C^k(M,S^1)/\varkappa_M C^k(M)$ followed by B. We now collect some of the material developed in this section:

Proposition 11 For any $k=o,\ldots,\infty$, the sequence

$$\underline{1} \longrightarrow \varkappa_M C^k(M) \xrightarrow{i} C^k(M,S^1) \xrightarrow{b} \Pi^1(M) \longrightarrow 1 \quad ,$$

in which i denotes the inclusion map and in which $\Pi^1(M)$ carries the discrete topology, is a topological exact sequence.

5.2. The character group of $C^k(M,S^1)$

Proposition 11 yields immediately that

$$\underline{1} \longrightarrow \Gamma_c \Pi^1(M) \xrightarrow{*b} \Gamma_c C^k(M,S^1) \xrightarrow{*i} \Gamma_c \varkappa_M C^k(M) \quad \text{is exact.}$$

The maps $*b$ and $*i$ are defined by composing characters with b and i respectively. Moreover, $*i$ is surjective as we will later show. The techniques involved are based on the universal covering $\tilde{M}$ of M and the subsequent lemmas. For their formulation, let us introduce $u:\tilde{M} \longrightarrow M$, the covering map of M and, for any k, its induced maps

$$u^* : C^k(M) \longrightarrow C^k(\tilde{M}),$$

which is a homeomorphism onto a subspace of $C^k(\tilde{M})$,

$$u^{**} : L_c C^k(\tilde{M}) \longrightarrow L_c C^k(M) \quad \text{and finally ,}$$

$$*u^* : \Gamma_c C^k(\tilde{M}) \longrightarrow \Gamma_c C^k(M) \ .$$

These maps are defined by composing functions in $C^k(M)$ with u, linear maps in $L_c C^k(\tilde{M})$ with u^* and characters in $\Gamma_c C^k(\tilde{M})$ with u^* respectively. By the Hahn-Banach theorem, u^{**} and hence $*u^*$ are surjective.

A convergence vector space E and a convergence group G will be called L_c-embeddable and P_c-embeddable if i_E and j_E are homeomorphisms onto subspaces of $L_c L_c E$ and $\Gamma_c \Gamma_c G$ respectively.

Lemma 12 For any $k=o,..,\infty$ both $L_c C^k(\tilde{M}) \xrightarrow{u^{**}} L_c C^k(M)$ and $\Gamma_c C^k(\tilde{M}) \xrightarrow{*u^*} \Gamma_c C^k(M)$ are quotient maps in the categories of L_c-embeddable convergence vector spaces and Γ_c-embeddable convergence groups respectively.

Proof: First let us prove the second assertion. Let G be a Γ_c-embeddable group and $h : \Gamma_c C^k(\tilde{M}) \longrightarrow G$ a continuous homomorphism which factors over $*u*$ to $\bar{h} : \Gamma_c C^k(M) \longrightarrow G$. The canonical maps $\varkappa_* \circ i^k_M$ and $\varkappa_* \circ i^k_{\tilde{M}}$ from M into $\Gamma_c C^k(M)$ and from $\tilde{M}$ into $\Gamma_c C^k(\tilde{M})$ are again denoted by the symbols j^k_M and $j^k_{\tilde{M}}$ respectively. For any $\gamma \in \Gamma G$ the map $\gamma \circ h \circ j^k_{\tilde{M}}$ is of class C^k and factors over u to the C^k-function $\gamma \circ \bar{h} \circ j^k_M$. Hence $*\bar{h}(\gamma) = \gamma \circ \bar{h} \in \Gamma_c \Gamma_c C^k(M)$. Since $*h = **u* \ *\bar{h}$, the map $*\bar{h}$ is continuous, and since G is P_c-embeddable h is continuous. The first assertion is verified analogously.

The next lemma employs $*u : C^k(M,S^1) \longrightarrow C^k(\tilde{M},S^1)$, defined by $t \circ u$ for each $t \in C^k(M,S^1)$. Since $\tilde{M}$ is simply connected, we have $C^k(\tilde{M},S^1) = \varkappa_{\tilde{M}} C^k(\tilde{M})$ for any k. Restricting $*u$ to $\varkappa_M C^k(M)$, we obtain the continuous homomorphism $**u : \Gamma_c \varkappa_M C^k(M) \longrightarrow \Gamma_c \varkappa_{\tilde{M}} C^k(\tilde{M})$, defined by composing the characters with $*u | \varkappa_M C^k(M)$.

Lemma 13 The homomorphism $\Gamma_c \varkappa_{\tilde{M}} C^k(\tilde{M}) \xrightarrow{**u} \Gamma_c \varkappa_M C^k(M)$ is a quotient map in the category of P_c-embeddable convergence groups.

Proof: In order to prove surjectivity, let us consider

$$\begin{array}{ccc} \Gamma_c \varkappa_M C^k(\tilde{M}) & \xrightarrow{*\varkappa_{\tilde{M}}} & \Gamma_c C^k(\tilde{M}) \\ \downarrow{**u} & & \downarrow{*u*} \\ \Gamma_c \varkappa_M C^k(M) & \xrightarrow{*\varkappa_M} & \Gamma_c C^k(M) \end{array},$$

for each $k=o,\ldots,\infty$, where $*\varkappa_M$ and $*\varkappa_{\tilde{M}}$ are defined in the usual way, namely by composing the character with $\varkappa_M$ and $\varkappa_{\tilde{M}}$ respectively. We first show that $**u$ is surjective. Consider $\gamma \in \Gamma_c \varkappa_M C^k(M)$ and form $*\varkappa_M(\gamma)$. By lemma 12 we can find a $\bar{\gamma} \in \Gamma_c C^k(\tilde{M})$ with $\bar{\gamma} \circ u* = *\varkappa_M(\gamma)$. Since $*\varkappa_M(\gamma)(\underline{\mathbb{Z}}) = \underline{1}$, and since $u*|\mathbb{Z} = id_{\mathbb{Z}}$, we have $\bar{\gamma}(\underline{\mathbb{Z}}) = \underline{1}$. Hence we find $\gamma^1 \in \Gamma_c \varkappa_M C^k(M)$ with $\gamma^1 \circ \varkappa_M = \bar{\gamma}$. Since $*\varkappa_M$ and $*\varkappa_{\tilde{M}}$ are injective, we have $**u(\gamma^1) = \gamma$. To verify the rest of the lemma, one proceeds analogously as in the proof of lemma 12, or one uses lemma 12 in connection with the direct decomposition of theorem 6.

We collect some of our results on the character group of $C^k(M,S^1)$ in the following theorem.

<u>Lemma 14</u> For any $k=o,\dots,\infty$ the topological exact sequences

$$\underline{0} \longrightarrow \underline{\mathbb{Z}} \xrightarrow{i_1} C^k(M) \xrightarrow{\varkappa_M} \varkappa_M C^k(M) \longrightarrow \underline{1}$$

and

$$\underline{1} \longrightarrow \varkappa_M C^k(M) \xrightarrow{i} C^k(M,S^1) \xrightarrow{b} \Pi^1(M) \longrightarrow 1$$

have exact duals, namely

$$\underline{1} \longrightarrow \Gamma_c \varkappa_M C^k(M) \xrightarrow{*\varkappa_M} \Gamma_c C^k(M) \xrightarrow{*i_1} \Gamma_c \underline{\mathbb{Z}} \cong S^1 \longrightarrow 1 ,$$

and

$$\underline{1} \longrightarrow \Gamma_c \Pi^1(M) \xrightarrow{*b} \Gamma_c C^k(M,S^1) \xrightarrow{*i} \Gamma_c \varkappa_M C^k(M) \longrightarrow \underline{1} .$$

Here $*\varkappa_M$ and $*b$ are homeomorphisms onto their ranges, $*i_1$ is a quotient map and $*i$ is a quotient map in the category of P_c-embeddable groups.

<u>Proof:</u> Since $\varkappa_M$ is a quotient map

$$\underline{1} \longrightarrow \Gamma \varkappa_M C^k(M) \xrightarrow{*\varkappa_M} \Gamma_c C^k(M) \xrightarrow{*i_1} \Gamma_c \mathbb{Z}$$

is right exact and $*\varkappa_M$ is a bicontinuous isomorphism onto a subspace. To show that the last map $*i_1$, which is a restriction map, is surjective, we extend a given character $\gamma \in \Gamma_c \mathbb{Z}$ onto $\mathbb{R}$, turn it via $\varkappa_*$ into a real-valued functional ℓ and extend this functional ℓ to $\ell' \in L_c C^k(M)$. Obviously, $\varkappa \circ \ell' = \gamma$. To show that $*i_1$, is a quotient map, one proceeds as in [Bi,2], p.71. To demonstrate that the second dual sequence is exact, we point out that the sequence

$$\underline{1} \longrightarrow \Gamma_c \Pi^1(M) \xrightarrow{*b} \Gamma_c C^k(M,S) \xrightarrow{*i} \Gamma_c \varkappa_M C^k(M)$$

is right exact, where $*b$ maps its domaine homeomorphically onto its range, regarded as a subspace of $\Gamma_c C^k(M,S^1)$.

To verify the surjectivity of $*i$, we form the commutative diagram:

$$\begin{array}{ccc} \Gamma_c \varkappa_{\tilde{M}} C^k(\tilde{M}) & \xrightarrow{**u} & \Gamma_c C^k(M,S^1) \\ \downarrow {\scriptstyle **u} & \swarrow {\scriptstyle *i} & \\ \Gamma_c \varkappa_M C^k(M) & & \end{array}$$

From this, we conclude via lemma 13, the last part of the above theorem.

5.3. The P_c-reflexivity of $C^k(M,S^1)$

For any $k=o,\dots,\infty$ consider the commuting diagram

$$\begin{array}{ccccccccc} \underline{1} & \longrightarrow & \Gamma_c\Gamma_c \varkappa_M C^k(M) & \xrightarrow{**i} & \Gamma_c\Gamma_c C^k(M,S^1) & \xrightarrow{**b} & \Gamma_c\Gamma_c \Pi^1(M) & \longrightarrow & \underline{1} \\ & & \uparrow {\scriptstyle j_{\varkappa_M C^k(M)}} & & \uparrow {\scriptstyle j_{C^k(M,S^1)}} & & \uparrow {\scriptstyle j_{\Pi^1(M)}} & & \\ \underline{1} & \longrightarrow & \varkappa_M C^k(M) & \xrightarrow{i} & C^k(M,S^1) & \xrightarrow{b} & \Pi^1(M) & \longrightarrow & 1 \end{array},$$

where $**i$ and $**b$ are defined by composing the respective characters with $*i$ and $*b$.

Both the discrete topological group $\Pi^1(M)$ and $\varkappa_M C^k(M)$ are P_c-reflexive (theorem 8). Using lemma 14, one easily verifies the exactness of the upper sequence. By the five lemma, $j_{C^k(M,S^1)}$ has to be an isomorphism. Evidently $j_{C^k(M,S^1)}$ is continuous. To see that its inverse is continuous we form $j_M^k : M \longrightarrow \Gamma_c C^k(M,S^1)$, defined by $j_M^k(p)(t)=t(p)$ for all $p \in M$ and all $t \in C^k(M,S^1)$. The dual map

$$*j_M^k : \Gamma_c\Gamma_c C^k(M,S^1) \longrightarrow C^k(M,S^1),$$

sends each $\gamma \in \Gamma_c\Gamma_c C^k(M,S^1)$ into $*j_M^k(\gamma) = \gamma \circ j_M^k$. Since $*j_M^k \circ j_{C^k(M,S^1)} = id_{C^k(M,S^1)}$, we obtain the continuity of $j^{-1}_{C^k(M,S^1)}$.

Therefore we may conclude with:

<u>Theorem 15</u> For any $k=o,\dots,\infty$ the topological group $C^k(M,S^1)$ is P_c-reflexive.

References

[Bi] E.Binz "Continuous Convergence on C(X)". Lecture Notes in Mathematics,Vol.469,1975, Springer-Verlag, Berlin, Heidelberg, New York

[Bi,2] E.Binz "Charaktergruppen von Gruppen von S^1-wertigen stetigen Funktionen",in Categorical Topology, Lecture Notes in Mathematics,Vol.540,1975, 43-92, Springer-Verlag,Berlin,Heidelberg,N.Y.

[Bi,Ke] E.Binz and H.H.Keller "Funktionenräume in der Kategorie der Limesräume". Ann.Acad.Sci.Fenn.Ser.A I, 383, 1966, 1-21.

[Bi,Bu,Ku] E.Binz, H.P.Butzmann, K.Kutzler "Über den c-Dual eines topologischen Vektorraumes". Math.Z., 127,1972,70-74.

[Bu] H.P.Butzmann "Pontryagin Dualität für topologische Vektorräume" (to appear in Arch.Math.)

[F-S] A.Freundlich-Smith "The Pontryagin Duality Theorem in Linear Spaces", Ann.Math., Vol.56,1952,248-253.

[Go,Gui] M.Golubitsky, V.Guillemin "Stable Mappings and their Singularities, Graduate Texts in Mathematics, Vol.14,1973, Springer-Verlag, New York, Berlin, Heidelberg

[Hu] S.T.Hu "Homotopy Theory",1959,Academic Press, New York, London.

[Ja] H.Jarchow "Duale Charakterisierung der Schwartz-Räume", Math.Ann,196,1972,85-90.

[Ke] H.H.Keller "Räume stetiger multilinearer Abbildungen als Limesräume",Math.Ann.Vol.159,1965, 259-270.

[Pa] R.S.Palais "Foundations of Global Non-Linear Analysis", 1968, W.A.Benjamin, Inc.,New York,Amsterdam.

[Po] L.S.Pontryagin "Topological Groups" (2nd Edition),1966, Gordon and Breach Science Pub.Inc.,New York, London,Paris.

[Schae] H.H.Schaefer "Topological Vector Spaces". Graduate Texts in Mathematics,Vol.3,1970,Springer-Verlag, New York,Berlin,Heidelberg.

[Spa] E.H.Spanier "Algebraic Topology",1966,Mc Graw-Hill Book Company,New York,London.

ENTROPY NUMBERS OF OPERATORS IN BANACH SPACES

B. Carl and A. Pietsch

Jena

In the following for every operator T between Banach spaces we define a sequence of so-called outer entropy numbers $e_n(T)$ with $n = 1, 2, \ldots$. Roughly speaking the asymptotic behaviour of $e_n(T)$ characterizes the "compactness" of T . In particular, T is compact if and only if $\lim_n e_n(T) = 0$.

The main purpose of this paper is to investigate the ideal $\mathcal{E}_p$ of all operators T such that

$$\sum_1^\infty e_n(T)^p < \infty .$$

For practical reason it is useful to introduce also inner entropy numbers $f_n(T)$ which, however, generate the same ideals.

The concept of entropy numbers is related to that of ε-entropy first studied by L. S. Pontrjagin and L. G. Schnirelman [13] in 1932. Further contributions are mainly due to Soviet mathematicians [1],[2]. For more information the reader is referred to the monograph of G. G. Lorentz [5], see also [4].

The significance of entropy numbers for the theory of operator ideals was discovered by the second named author. A full account will be given in [12].

In the following E , F and G are real Banach spaces. The closed unit ball of E is denoted by U_E . Furthermore, $\mathcal{L}(E, F)$ denotes the Banach space of all (bounded and linear) operators from E into F . The symbols l_p^n and l_p stand for the classical Banach spaces of vectors and sequences, respectively.

All logarithms are to the base 2 .

1. Elementary properties of entropy numbers

For every operator $T \in \mathcal{L}(E, F)$ the n-th <u>outer entropy number</u> $e_n(T)$ is defined to be the infimum of all $\sigma \geqq 0$ such that there are $y_1, \ldots, y_q \in F$ with $q \leqq 2^{n-1}$ and

$$T(U_E) \subseteq \bigcup_1^q \{y_i + \sigma U_F\} .$$

For every operator $T \in \mathcal{L}(E, F)$ the n-th <u>inner entropy number</u> $f_n(T)$ is defined to be the supremum of all $\rho \geqq 0$ such that there are $x_1, \ldots, x_p \in U_E$ with $p > 2^{n-1}$ and

$$\|Tx_i - Tx_k\| > 2\rho \qquad \text{for } i \neq k .$$

First we state an elementary property of entropy numbers.

Proposition 1.

If $T \in \mathcal{L}(E, F)$, then

$$\|T\| = e_1(T) \geqq e_2(T) \geqq \ldots \geqq 0 \quad \text{and} \quad \|T\| = f_1(T) \geqq f_2(T) \geqq \ldots \geqq 0 .$$

Next we check the so-called additivity of entropy numbers.

Proposition 2.

If $T_1, T_2 \in \mathcal{L}(E, F)$, then

$$e_{n_1+n_2-1}(T_1 + T_2) \leqq e_{n_1}(T_1) + e_{n_2}(T_2)$$

and

$$f_{n_1+n_2-1}(T_1 + T_2) \leqq f_{n_1}(T_1) + f_{n_2}(T_2) .$$

Proof.

Let $T_1, T_2 \in \mathcal{L}(E, F)$. If $\sigma_k > e_{n_k}(T_k)$, then there are $y_1^{(k)}, \ldots, y_{q_k}^{(k)} \in F$ such that

$$T_k(U_E) \subseteq \bigcup_{i=1}^{q_k} \left\{y_i^{(k)} + \sigma_k U_F\right\} \quad \text{and} \quad q_k \leqq 2^{n_k-1} \quad \text{for } k = 1,2.$$

Hence, given $x \in U_E$, we can find i_k and $y_k \in U_F$ with

$$T_k x = y_{i_k}^{(k)} + \sigma_k y_k \qquad \text{for } k = 1, 2 .$$

It follows from

$$(T_1+T_2)x \in y_{i_1}^{(1)} + y_{i_2}^{(2)} + (\sigma_1 + \sigma_2)U_F$$

that

$$(T_1+T_2)(U_E) \subseteq \bigcup_{i_1=1}^{q_1} \bigcup_{i_2=1}^{q_2} \left\{y_{i_1}^{(1)} + y_{i_2}^{(2)} + (\sigma_1 + \sigma_2) U_F\right\} .$$

Since $q_1 q_2 \leqq 2^{(n_1+n_2-1)-1}$, we get $e_{n_1+n_2-1}(T_1+T_2) \leqq \sigma_1 + \sigma_2$.

This shows the desired inequality for outer entropy numbers. The remaining part of the proof is left to the reader.

The multiplicativity of entropy numbers can be proved with the same method.

Proposition 3.

If $T \in \mathcal{L}(E, F)$ and $S \in \mathcal{L}(F, G)$, then

$$e_{m+n-1}(ST) \leqq e_m(S)\, e_n(T)$$

and

$$f_{m+n-1}(ST) \leqq f_m(S)\, f_n(T).$$

Finally, the relationship between outer and inner entropy numbers is investigated.

Proposition 4.

If $T \in \mathcal{L}(E, F)$, then

$$f_n(T) \leqq e_n(T) \leqq 2\, f_n(T) .$$

Proof.

Suppose that $\sigma > e_n(T)$ and $\rho < f_n(T)$. Then we can find $x_1, \ldots, x_p \in U_E$ and $y_1, \ldots, y_q \in F$ with $\|Tx_i - Tx_j\| > 2\rho$ for $i \neq j$ and $T(U_E) \subset \bigcup_{k=1}^{q} \{y_k + \sigma U_F\}$, where $p > 2^{n-1} \geqq q$. So there must exist different elements Tx_i and Tx_j which belong to the same set $y_k + \sigma U_F$. Consequently $2\rho < \|Tx_i - Tx_j\| \leqq 2\sigma$. This proves that $f_n(T) \leqq e_n(T)$. Given $\rho > f_n(T)$, we choose a maximal family of elements $x_1, \ldots, x_p \in U_E$ such that $\|Tx_i - Tx_k\| > 2\rho$ for $i \neq K$. Clearly $p \leqq 2^{n-1}$. Moreover, for $x \in U_E$ we can find some i with $\|Tx - Tx_i\| \leqq 2\rho$. This means that

$$T(U_E) \subseteq \bigcup_{1}^{p} \{Tx_i + 2\rho\, U_F\} .$$

So $e_n(T) \leqq 2\rho$ and therefore $e_n(T) \leqq 2\, f_n(T)$.

2. Quasi-normed operator ideals related to entropy numbers

In the following let $\mathcal{L}$ denote the class of all operators between Banach spaces while $\mathcal{K}$ denotes the closed ideal of compact operators. Then we have

$$\mathcal{K} = \{T \in \mathcal{L} : (e_n(T)) \in c_o\} .$$

Therefore it seems very natural to introduce the following class of operators. Given $0 < p < \infty$, we define

$$\mathscr{E}_p := \{T \in \mathscr{L} : (e_n(T)) \in l_p\} .$$

Moreover, for $T \in \mathscr{E}_p$ we put

$$E_p(T) := (\sum_1^{\infty} e_n(T)^p)^{1/p} .$$

We now show that $\mathscr{E}_p$ is a so-called operator ideal for which every component $\mathscr{E}_p(E, F)$ becomes a complete metric linear space with respect to the quasi-norm E_p .

Theorem 1.

If $T_1, T_2 \in \mathscr{E}_p(E, F)$, then $T_1 + T_2 \in \mathscr{E}_p(E, F)$ and

$$E_p(T_1 + T_2) \leqq c\,[E_p(T_1) + E_p(T_2)] ,$$

where

$$c := 2^{1/p} \max (2^{1/p-1}, 1) .$$

Proof.

By Proposition 1 and 2 we get

$$\begin{aligned} E_p(T_1 + T_2) &= \Big\{\sum_1^{\infty} e_n(T_1 + T_2)^p\Big\}^{1/p} \\ &\leqq \Big\{2 \sum_1^{\infty} e_{2n-1}(T_1 + T_2)^p\Big\}^{1/p} \\ &\leqq 2^{1/p} \Big\{\sum_1^{\infty} [e_n(T_1) + e_n(T_2)]^p\Big\}^{1/p} \\ &\leqq c\,[E_p(T_1) + E_p(T_2)] . \end{aligned}$$

Remark.

It follows from Proposition 4 that

$$\mathscr{E}_p = \{T \in \mathscr{L} : (f_n(T)) \in l_p\} .$$

Moreover, by setting

$$F_p(T) := (\sum_1^{\infty} f_n(T)^p)^{1/p}$$

we define a quasi-norm F_p equivalent to E_p .

Without proof we state

Theorem 2.

If $X \in \mathscr{L}(E_o, E)$, $T \in \mathscr{E}_p(E, F)$ and $Y \in \mathscr{L}(F, F_o)$, then $YTX \in \mathscr{E}_p(E_o, F_o)$ and $E_p(YTX) \leqq \|Y\| \; E_p(T) \; \|X\|$.

The following statement is also evident.

Proposition 5.

If $0 < p_1 < p_2 < \infty$, then $\mathscr{E}_{p_1} \subset \mathscr{E}_{p_2}$ and the embedding map is continuous.

Theorem 3.

If $0 < p, q < \infty$ and $\frac{1}{r} = \frac{1}{p} + \frac{1}{q}$, then $T \in \mathscr{E}_q(E, F)$ and $S \in \mathscr{E}_p(F, G)$ imply that $ST \in \mathscr{E}_r(E, G)$ and $E_r(ST) \leqq$ $\leqq 2^{1/r} E_p(S) \, E_q(T)$.

Proof.

By Proposition 1 and 3 we get

$$\begin{aligned} E_r(ST) &= \left\{ \sum_1^{\infty} e_n(ST)^r \right\}^{1/r} \\ &\leqq \left\{ 2 \sum_1^{\infty} e_{2n-1}(ST)^r \right\}^{1/r} \\ &\leqq 2^{1/r} \left\{ \sum_1^{\infty} \left[e_n(S) \, e_n(T) \right]^r \right\}^{1/r} \\ &\leqq 2^{1/r} E_p(S) \, E_q(T) . \end{aligned}$$

This proves the assertion.

3. Quasi-normed operator ideals related to approximation numbers

For every operator $T \in \mathscr{L}(E, F)$ the n-th approximation number is defined by

$$a_n(T) := \inf \left\{ \|T - L\| : L \in \mathscr{L}(E, F) \text{ and rank } (L) < n \right\}.$$

As shown in [9] or [10] the class

$$\mathscr{S}_p := \left\{ T \in \mathscr{L} : \sum_1^{\infty} a_n(T)^p < \infty \right\}, \; 0 < p < \infty ,$$

is an operator ideal for which every component $\mathscr{S}_p(E, F)$ becomes a complete metric linear space with respect to the quasi-norm

$$S_p(T) := \left(\sum_1^\infty a_n(T)^p\right)^{1/p} .$$

Only a little is known about the relationship between $\mathscr{E}_p$ and $\mathscr{S}_p$.

Conjecture 1.

If $0 < p < \infty$, then $\mathscr{S}_p \subseteq \mathscr{E}_p$.

Conjecture 2.

If $0 < p < 2$ and $\frac{1}{q} = \frac{1}{p} - \frac{1}{2}$, then $\mathscr{E}_p \subseteq \mathscr{S}_q$.

The inclusions stated above are the best possible which can be expected. Some weaker results are proved in [12].

4. Entropy numbers of operators in Hilbert spaces

It seems to be very complicated to compute or estimate the entropy numbers of a given operator. However, we know some results concerning the related quasi-norms.

Theorem 4.

Let $S \in \mathscr{L}(l_2, l_2)$ such that $S(\xi_n) = (\sigma_n \xi_n)$ and $(\sigma_n) \in c_o$. If $\sigma_1 \geqq \sigma_2 \geqq \dots \geqq 0$, then

$$\sigma_n \leqq 2e_n(S) .$$

Proof.

If $\sigma_n = 0$, then the assertion is trivial. So we assume that $\sigma_1 \geqq \sigma_2 \geqq \dots \geqq \sigma_n > 0$. Put

$$J_n(\xi_1, \dots, \xi_n) := (\xi_1, \dots, \xi_n, 0, \dots)$$

and

$$Q_n(\xi_1, \dots, \xi_n, \xi_{n+1}, \dots) := (\xi_1, \dots, \xi_n) .$$

Then $S_n = Q_n S J_n$ is invertible. If I_n denotes the identity map of l_2^n , it follows from $e_n(I_n) \geqq 1/2$ and Proposition 2 that

$$1/2 \leqq e_n(I_n) \leqq e_n(S_n) \, \|S_n^{-1}\| \leqq \|Q_n\| \, e_n(S) \, \|J_n\| \, \sigma_n^{-1}$$

$$\leq e_n(S)\, \sigma_n^{-1} .$$

This completes the proof.

Theorem 5.

Let $S \in \mathcal{L}(l_2, l_2)$ such that $S(\xi_n) = (\sigma_n \xi_n)$ and $(\sigma_n) \in c_o$. Then

$$\Big(\sum_1^{\infty} e_n(S)^p\Big)^{1/p} \leq c_p\Big(\sum_1^{\infty} |\sigma_n|^p\Big)^{1/p} \quad \text{for } 0 < p < \infty ,$$

where c_p is some positive constant.

Proof.

Without loss of generality we may suppose that $\sigma_1 \geq \sigma_2 \geq \dots \geq 0$. Let

$$E(\varepsilon) := \max \{n : e_n(S) > \varepsilon\} \qquad \text{for } 0 < \varepsilon < \sigma_1 .$$

We now show that

$$(*) \qquad E(2\varepsilon) \leq 1 + \sum_{\sigma_k > \varepsilon} \log (8\, \sigma_k/\varepsilon) .$$

Put $m := \max (k : \sigma_k > \varepsilon)$ and $S_m := Q_m S J_m$. Let U_2^m and U_∞^m denote the closed unit ball of l_2^m and l_∞^m, respectively. If $y \in S_m(U_2^m)$, then there exists $g = (\gamma_1, \dots, \gamma_m)$ such that

$$y \in \varepsilon m^{-1/2} \{2g + U_\infty^m\} \subseteq 2\varepsilon\, m^{-1/2} g + \varepsilon U_2^m ,$$

where $\gamma_1, \dots, \gamma_m$ are integers. Since $\sigma_1 \geq \dots \geq \sigma_m > \varepsilon$, we have

$$\varepsilon m^{-1/2} \{2g + U_\infty^m\} \subseteq S_m(U_2^m) + 2\varepsilon\, m^{-1/2}\, U_\infty^m \subseteq 3S_m(U_2^m) .$$

Let $g_1, \dots, g_q$ be the collection of all $g_i = (\gamma_{i1}, \dots, \gamma_{im})$ with

$$\varepsilon m^{-1/2} \{2g_i + U_\infty^m\} \subseteq 3\, S_m(U_2^m) .$$

Clearly

$$S_m(U_2^m) \subseteq \bigcup_1^q \{2\varepsilon\, m^{-1/2} g_i + \varepsilon U_2^m\}$$

and therefore

$$S(U_2) \subseteq J_m\, S_m(U_2^m) + \sigma_{m+1} U_2 \subseteq \bigcup_1^q \{2\varepsilon m^{-1/2}\, J_m g_i + 2\varepsilon U_2\},$$

where U_2 denotes the closed unit ball of l_2 . On the other hand,

$$q\, [\varepsilon m^{-1/2}]^m\, \lambda(U_\infty^m) = \sum_1^q \lambda[\varepsilon m^{-1/2}\{2g_i + U_\infty^m\}] \leq 3^m \prod_1^m \sigma_k\, \lambda(U_2^m),$$

where λ is the Lebesgue measure. Using Stirling's formula we get

$$e^t\Gamma(t+1) \geqq \sqrt{2\pi}\, t^{t+\frac{1}{2}} \quad \text{for} \quad 0 < t < \infty .$$

Hence

$$\lambda(U_2^m) = \frac{\pi^{\frac{m}{2}}}{\Gamma(\frac{m}{2}+1)} \leqq \frac{(2\pi e)^{\frac{m}{2}}}{m^{m/2}} \leqq \frac{5^m}{m^{m/2}} .$$

This implies that

$$q\,\varepsilon^m \leqq 8^m \prod_1^m \sigma_k .$$

Choose n such that

$$n-1 \leqq 1 + \sum_1^m \log(8\,\sigma_k/\varepsilon) \leqq n .$$

Then $q \leqq 2^{n-1}$ and therefore $e_n(S) \leqq 2\varepsilon$. So $E(2\varepsilon) \leqq n-1$. This proves ($*$).

Finally, we have

$$\begin{aligned} 2^{-p}\sum_1^\infty e_n(S)^p &= 2^{-p}\sum_1^\infty n\left[e_n(S)^p - e_{n+1}(S)^p\right] \\ &= 2^{-p}\int_0^{\sigma_1} E(\varepsilon)\, d\varepsilon^p \\ &\leqq \sigma_1^p + \int_0^{\sigma_1} \sum_{\sigma_k>\varepsilon} \log(8\,\sigma_k/\varepsilon)\, d\varepsilon^p \\ &= \sigma_1^p + \sum_{i=1}^\infty \int_{\sigma_{i+1}}^{\sigma_i} \sum_{\sigma_k>\varepsilon} \log(8\,\sigma_k/\varepsilon)\, d\varepsilon^p \\ &= \sigma_1^p + \sum_{i=1}^\infty \sum_{k=1}^i \int_{\sigma_{i+1}}^{\sigma_i} \log(8\,\sigma_k/\varepsilon)\, d\varepsilon^p \\ &= \sigma_1^p + \sum_{k=1}^\infty \sum_{i=k}^\infty \int_{\sigma_{i+1}}^{\sigma_i} \log(8\,\sigma_k/\varepsilon)\, d\varepsilon^p \\ &= \sigma_1^p + \sum_{k=1}^\infty \int_0^{\sigma_k} \log(8\,\sigma_k/\varepsilon)\, d\varepsilon^p \\ &= \sigma_1^p + \frac{8^p}{p}\int_0^{8^{-p}} \log(1/t)\,dt \sum_1^\infty \sigma_k^p . \end{aligned}$$

This completes the proof.

The above theorems show that for any Hilbert space H the operator ideal $\mathscr{C}_p(H, H)$ coincides with the operator ideal $\mathscr{S}_p(H,H)$ In particular, $\mathscr{C}_2(H, H)$ is the ideal of so-called Hilbert-Schmidt Operators.

5. Entropy quasi-norms of the identity map I_n from l_u^n to l_v^n

Lemma 1.

If $m = 1, \ldots, n$, then

$$e_m(I_n : l_\infty^n \to l_1^n) \geq \frac{1}{2e} n .$$

Proof.

Let U_∞^n and U_1^n denote the closed unit ball of l_∞^n and l_1^n, respectively. Suppose that

$$U_\infty^n \subseteq \bigcup_1^q \{y_i + \sigma U_1^n\} \quad \text{and} \quad q \leq 2^{n-1} .$$

Then

$$\lambda(U_\infty^n) \leq \sum_1^q \lambda(y_i + \sigma U_1^n) = q\, \sigma^n \lambda(U_1^n) ,$$

where λ is the Lebesgue measure on R^n. Now $\lambda(U_\infty^n) = 2^n$ and $\lambda(U_1^n) = 2^n/n!$ imply that $\sigma^n \geq n!/2^{n-1}$. Using $e^n n! > n^n$ we get $\sigma > n/2e$. Therefore

$$e_n(I_n : l_\infty^n \to l_1^n) \geq n/2e .$$

In order to prove the following lemma we use a decomposition-trick taken from M. Š. Birman and M. Z. Solomjak [1].

Lemma 2.

If $m = 1, \ldots, n$, then

$$e_m(I_n : l_1^n \to l_\infty^n) \leq c\, \frac{\log(n+1)}{m} ,$$

where c is a positive constant.

Proof.

Let U_1^n and U_∞^n be as before. If $m \geq 4$, then

$$\sigma := 4\, \frac{\log(n+1)}{m} \geq 2\, \frac{\log(n+1)}{m-2} > \frac{1}{n} .$$

Put

$$K(x) := \{ k : |\xi_k| > \sigma \} \quad \text{for} \quad x = (\xi_k) \in U_1^n .$$

We have

$$\text{card}\,(K(x)) < \sum_{K(x)} \frac{|\xi_k|}{\sigma} \leq 1/\sigma < n .$$

Let $\mathbb{K}$ denote the collection of all sets $K \subseteq \{1, \ldots, n\}$ with $\operatorname{card}(K) < 1/\sigma$ and put

$$U_K := \{x \in U_\infty^n : \xi_k = 0 \text{ if } k \notin K\}.$$

Then

$$x \in U_{K(x)} + \sigma U_\infty^n \quad \text{for all } x \in U_1^n .$$

Hence

$$U_1^n \subseteq \bigcup_{\mathbb{K}} \{U_K + \sigma U_\infty^n\}.$$

Clearly, we can find $y_i^{(K)} \in l_\infty^n$ such that

$$U_K \subseteq \bigcup_1^{q_K} \{y_i^{(K)} + \sigma U_\infty^n\} \quad \text{and} \quad q_K \leqq (1/\sigma + 1)^{\operatorname{card}(K)} .$$

Consequently, there are $y_i \in l_\infty^n$ with

$$U_1^n \subseteq \bigcup_1^q \{y_i + 2\sigma U_\infty^n\}$$

and

$$q \leqq \sum_{\mathbb{K}} (1/\sigma + 1)^{\operatorname{card}(K)} \leqq \sum_1^{1/\sigma} \binom{n}{h} (1/\sigma + 1)^h \leqq 2(n+1)^{2/\sigma} \leqq 2^{m-1}.$$

So we get

$$e_m(I_n : l_1^n \to l_\infty^n) \leqq 2\sigma \leqq 8 \frac{\log(n+1)}{m} .$$

Obviously this estimate is also true for $m = 1, 2, 3$.

Proposition 6.

If $0 < p < \infty$, then

$$E_p(I_n : l_\infty^n \to l_1^n) \geqq a_p \, n^{1/p+1} \quad \text{for } n = 1, 2, \ldots,$$

where a_p is some positive constant.

Proof.

By Lemma 1 we have

$$e_m(I_n : l_\infty^n \to l_1^n) \geqq \frac{1}{2e} n \quad \text{for } m = 1, \ldots, n .$$

Therefore

$$E_p(I_n : l_\infty^n \to l_1^n) \geqq \frac{1}{2e} n^{1/p+1} .$$

Proposition 7.

If $0 < p < 1$, then

$$E_p(I_n : l_1^n \to l_\infty^n) \leqq b_p\, n^{1/p-1} \log(n+1) \quad \text{for} \quad n = 1, 2, \ldots,$$

where b_p is some positive constant.

Proof.

Using Proposition 3 we have

$$E_p(I_n : l_1^n \to l_\infty^n) \leqq \Big(\sum_{k=0}^{\infty} \sum_{m=1}^{n} e_{kn+m}(I_n : l_1^n \to l_\infty^n)^p\Big)^{1/p}$$

$$\leqq \Big(\sum_{m=1}^{n} e_m(I_n : l_1^n \to l_\infty^n)^p\Big)^{1/p} \Big(\sum_{k=0}^{\infty} e_{n+1}(I_n : l_\infty^n \to l_\infty^n)^{kp}\Big)^{1/p}.$$

From $e_{n+1}(I_n : l_\infty^n \to l_\infty^n) = 1/2$ we get

$$\Big(\sum_{k=0}^{\infty} e_{n+1}(I_n : l_\infty^n \to l_\infty^n)^{kp}\Big)^{1/p} \leqq c_p \,.$$

By Lemma 2 it follows that

$$\Big(\sum_{m=1}^{n} e_m(I_n : l_1^n \to l_\infty^n)^p\Big)^{1/p} \leqq d_p\, n^{1/p-1} \log(n+1) \,.$$

Since the constants c_p and d_p do not depend on n, the assertion is proved.

Theorem 6.

If $0 < p < 1$ and $1 \leqq u, v \leqq \infty$, then

$$a_p n^{1/p+1/v-1/u} \leqq E_p(I_n : l_u^n \to l_v^n) \leqq b_p n^{1/p+1/v-1/u} \log(n+1)$$

$$\text{for} \quad n = 1, 2, \ldots,$$

where a_p and b_p are positive constants.

Proof.

By Theorem 2 and Proposition 6 we get

$$a_p n^{1/p+1} \leqq E_p(I_n : l_\infty^n \to l_1^n) \leqq \|I_n : l_\infty^n \to l_u^n\| E_p(I_n : l_u^n \to l_v^n) \, \|I_n : l_v^n \to l_1^n\|$$

$$\leqq n^{1/u} E_p(I_n : l_u^n \to l_v^n)\, n^{1-1/v}$$

and therefore

$$a_p\, n^{1/p+1/v-1/u} \leqq E_p(I_n : l_u^n \to l_v^n) \,.$$

Analogously, by Theorem 2 and Proposition 7 we have

$$E_p(I_n : l_u^n \to l_v^n) \leqq \|I_n : l_u^n \to l_1^n\| \; E_p(I_n : l_1^n \to l_\infty^n) \, \|I_n : l_\infty^n \to l_v^n\|$$

$$\leq n^{1-1/u} b_p n^{1/p-1} \log(n+1) n^{1/v} =$$

$$= b_p n^{1/p+1/v-1/u} \log(n+1) .$$

The limit order $\lambda(\mathscr{E}_p, u, v)$ is defined to be the infimum of all $\lambda \geq 0$ such that

$$E_p(I_n: l_u^n \rightarrow l_v^n) \leq c\, n^{\lambda} \quad \text{for } n = 1, 2, \dots ,$$

where c is some constant. Using this concept we can restate the above result as follows.

Theorem 7.

If $0 < p < 1$ and $1 \leq u, v \leq \infty$, then

$$\lambda(E_p, u, v) = 1/p+1/v-1/u .$$

The remaining case is treated in the next theorem. For the proof the reader is referred to [12].

Theorem 8.

If $1 \leq p < \infty$ and $1 \leq u, v \leq \infty$, then

$$\lambda(E_p, u, v) = \max(1/p+1/v-1/u, 0).$$

The limit order is very useful for formulating conditions for a given diagonal operator $S(\xi_n) = (\sigma_n \xi_n)$ to belong to $\mathscr{E}_p(l_u, l_v)$. According to a deep theorem of H. König (3) our results can also be carried across to embedding maps of Sobolev spaces and to weakly singular integral operators from L_u into L_v.

References

[1] M. Š. Birman and M. Z. Solomjak: Piecewise polynomial approximations of functions of classes W_p. (Russian) Mat. Sb. 73 (115) (1967), 331-355.

[2] A. N. Kolmogorov and W. M. Tichomirow: ε-entropy and ε-capacity of sets in function spaces. (Russian) Uspehi Mat. Nauk 14 (86) (1959), 3-86. German translation in Arbeiten zur Informationstheorie III, Berlin 1960.

[3] H. König: Grenzordnungen von Operatorenidealen I, II. Math. Ann. 212 (1974), 51-77.

[4] G. G. Lorentz: Metric entropy and approximation. Bull. Amer. Math. Soc. 72 (1966), 903-937.

[5] G. G. Lorentz: Approximation of functions. New York, 1966.

[6] B. S. Mitjagin: Approximative dimension and bases in nuclear spaces. (Russian) Uspehi Mat. Nauk 16 (4) (1961), 63-132.

[7] B. S. Mitjagin and A. Pełczyński: Nuclear operators and approximative dimension. Proc. of ICM, Moscow 1966, 366-372.

[8] R. Oloff: Entropieeigenschaften von Diagonaloperatoren. (To appear.)

[9] A. Pietsch: Einige neue Klassen von kompakten linearen Abbildungen. Rev. Roumaine Math. Pures Appl. 8 (1963), 427-447.

[10] A. Pietsch: Nuclear locally convex spaces. Berlin-Heidelberg-New York, 1972.

[11] A. Pietsch: s-numbers of operators in Banach spaces. Studia Math. 51 (1974), 201-223.

[12] A. Pietsch: Operator ideals. Berlin, 1978.

[13] L. S. Pontrjagin and L. G. Schnirelman: Sur une propriété métrique de la dimension. Ann. of Math. 33 (1932), 156-162.

SOME RECENT APPLICATIONS OF ULTRAFILTERS TO TOPOLOGY

W. W. Comfort
Wesleyan University
Middletown, Connecticut, U.S.A.

The three theorems given here hardly exhaust the applications of ultrafilters on ω to topology, of course - indeed, they barely scratch the surface. I have selected them for discussion at this symposium because (in my judgement) they are unusually elegant and pretty results and because they share in common this feature: neither the hypotheses nor the conclusion of these theorems mention ultrafilters, but ultrafilters on ω play a crucial rôle in each of the proofs given here.

These proofs are due to the workers who are identified below. They have circulated informally among aficionadoes, but they have not yet appeared in printed form. I am grateful to these mathematicians for authorizing and encouraging both this brief exposition of their work and the more detailed treatment anticipated in my forthcoming account [5].

§1. *Notation and Terminology*. Throughout these remarks, by a *space* we mean completely regular, Hausdorff space. We denote by ω the least infinite cardinal. The symbol α denotes an (arbitrary) infinite cardinal and the discrete topological space of cardinality α. For X a space we denote by βX the Stone-Čech compactification of X. As is well-known (see for example [12], [6]), $\beta(\alpha)$ may be identified with the space of ultrafilters on α topologized so that $\{\{p \in \beta(\alpha) : A \in p\} : A \subset \alpha\}$ is a base for the closed sets; the inclusion $\alpha \subset \beta(\alpha)$ is effected by identifying the element ξ of α with the principal ultrafilter $\{A \subset \alpha : \xi \in A\}$.

If X and Y are spaces and f is a continuous function from X into Y, we denote by $\overline{f}$ that (unique) continuous function from βX to βY such that $f \subset \overline{f}$. (It may be argued that since $f[X] \subset Y$ for many spaces Y, the function $\overline{f}$ is not well-defined. We hope that in each case our intention concerning Y will be clear; in most cases Y will be taken compact, so that $\overline{f}[X] \subset Y$.)

For $\alpha \geq \omega$ we define

$$U(\alpha) = \{p \in \beta(\alpha) : |A| = \alpha \text{ for all } A \in p\},$$

and we note that $U(\omega) = \beta(\omega) \sim \omega$.

§2. *A Theorem of Frolík and Kunen.*

A subset A of a space X is said to be C^*-*embedded* in X if for every continuous function from A to the space $[0,1]$ (equivalently: to a compact space) there is continuous function g on X such that $f \subset g$.

We state the theorem of Frolík and Kunen.

2.1. *Theorem.* Let X be an infinite compact space in which each infinite, discrete subspace is C^*-embedded. Then X is not homogeneous.

Remarks Concerning the Proof. Suppose we can show that there are $p, q \in U(\omega)$ such that for every $f \in \omega^\omega$ we have $\overline{f}(p) \neq q$ and $\overline{f}(q) \neq p$. Arranging the notation so that $\beta(\omega) \subset X$, we claim that if h is a homeomorphism of X onto X then $h(p) \neq q$. Indeed if $h(p) = q$ then one of these four events occurs: $q \in (h[\omega] \cap \omega)^-$; $q \in (h[\omega] \cap U(\omega))^-$; $p \in (h^{-1}[\omega] \cap \omega)^-$; $p \in (h^{-1}[\omega] \cap U(\omega))^-$. In the first case there is $f \in \omega^\omega$ so that (f agrees with h on $h^{-1}[h[\omega] \cap \omega] \in p$ and) $\overline{f}(p) = q$; in the second case there is a one-to-one function g from ω to $U(\omega)$ such that $g[\omega]$ is discrete and g agrees with h on $h^{-1}[h[\omega] \cap U(\omega)]$, and hence (as is easily shown) there is $f \in \omega^\omega$ such that $\overline{f}(q) = p$; the third case is similar to the first, and the fourth to the second.

To complete the argument, it remains to show that there are $p, q \in U(\omega)$ such that if $f \in \omega^\omega$ then $\overline{f}(p) \neq q$ and $\overline{f}(q) \neq p$. This is a recent result of Kunen [19], recorded also in [6] (Theorem 10.4) and [5], and we shall not repeat it here. Those portions of the argument outlined above had already been supplied by Frolík [8], [9], [10] en route to a number of elegant non-homogeneity results (including the statement that $U(\omega)$ is not homogeneous).

We record some consequences of Theorem 2.1. (A space is said to be *extremally disconnected* if the closure of each of its open subsets is open. A space is an F-*space* if each of its cozero-sets is C^*-embedded.)

2.2. *Corollary.* Let X be an infinite, compact space which satisfies one of the following conditions. Then X is not homogeneous.

(a) X is an F-space;

(b) there is an extremally disconnected space Y such that $X \subset Y$;

(c) there is $\alpha \geq \omega$ such that $X \subset \beta(\alpha)$;

(d) there is a locally compact, σ-compact space Y such that $X = \beta Y \setminus Y$.

Proof. (a) It is easy to show that every countable, discrete subspace of an F-space is C^*-embedded. It is known, more generally, that every countable subspace of an F-space is C^*-embedded (see [12] (Problem 14N) or [6] (Lemma 16.15(b))).

(b) βY is extremally disconnected, hence an F-space. Since X is C^*-embedded in βY, X is itself an F-space and hence (a) applies.

It is clear that (c) => (b).

Gillman and Henriksen [11] have shown that (d) => (a). An elegant proof, due to Negrepontis, is given in [6] (Lemma 14.16).

The proof of Corollary 2.2 is complete.

We note that Theorem 2.1 was announced (for compact, extremally disconnected spaces) in an editorial footnote appended to [7].

2.3. *A Question.* Let $\alpha \geq \omega$, and let $C(\alpha)$ denote the set of all cardinals γ for which there is $S \subset U(\alpha)$ such that

(1) $|S| = \gamma$, and

(2) if $p, q \in S$ and $p \neq q$, and if $f \in \alpha^\alpha$, then $\overline{f}(p) \neq q$ and $\overline{f}(q) \neq p$.

As indicated above, Kunen [19] has shown $2 \in C(\omega)$. In fact, Kunen has shown (without any special assumptions concerning any cardinal numbers) that $2^\alpha \in C(\alpha)$. This result suggests the following questions.

Is $(2^\alpha)^+ \in C(\alpha)$? Is $\sup C(\alpha) \in C(\alpha)$? Is $2^{2^\alpha} = \sup C(\alpha)$? Is $2^{2^\alpha} \in C(\alpha)$?

It is shown in [6] and [5] that the answers are affirmative for all α such that $\alpha^+ = 2^\alpha$ or $(2^\alpha)^+ < 2^{2^\alpha}$.

§ 3. *A Theorem of Ginsburg and Saks.*

We denote by $\{X_i : i \in I\}$ a set of non-empty spaces, for $\emptyset \neq J \subset I$ we write X_J in place of $\prod_{i \in J} X_i$, and we denote by π_J the projection from X_I onto X_J.

We say that a space X is *countably compact* provided that for every $f \in X^\omega$ there is $p \in U(\omega)$ such that $\overline{f}(p) \in X$. (In the context of our spaces, this definition agrees with other more usual definitions.)

3.1. *Theorem.* The space X_I is countably compact if and only if X_J is countably compact for all $J \subset I$ such that $0 < |J| \leq 2^{2^\omega}$.

Remarks Concerning the Proof. The "only if" implication follows from the fact that the continuous image of a countably compact space is countably compact.

We turn to the "if" implication.

Following Bernstein [2], we say (for $p \in U(\omega)$) that a space X is *p-compact* if $\overline{f}(p) \in X$ for every $f \in X^\omega$. It is not difficult to show (cf. [2] or [13]) that the product of p-compact spaces is p-compact. Indeed we have the following statement:

(*) If $f \in (X_I)^\omega$ and if there is $x = \langle x_i : i \in I \rangle \in X_I$ such that $(\pi_i \circ f)^-(p) = x_i$ for all $i \in I$, then $\overline{f}(p) = x$.

It follows that if the desired conclusion fails then there are $f \in (X_I)^\omega$, and (for every $p \in U(\omega)$) an element $i(p)$ of I, such that $(\pi_{i(p)} \circ f)^-(p) \notin X_{i(p)}$. Then with $J = \{i(p) : p \in U(\omega)\}$ we have $\pi_J \circ f \in (X_J)^\omega$ and $(\pi_J \circ f)^- [U(\omega)] \cap X_J = \emptyset$, so that X_J is not countably compact. Since $|J| \leq |U(\omega)| = 2^{2^\omega}$, the proof is complete.

We remark that (*) may be established directly, as in [5], or by appeal to the work of Glicksberg [15]. It is noted in [15] that a product space X_I is pseudocompact if and only if X_J is pseudocompact for all $J \subset I$ such that $0 < |J| \leq \omega$; thus our space X_I is pseudocompact. It then follows from Theorem 1 of [15] that $\beta(X_I) = \prod_{i \in I} \beta X_i$; statement (*) is then obvious.

The technique used in the proof just given has been used by Ginsburg and Saks [13] and Saks [20] in connection with product-space theorems concerning a multitude of topological properties.

Theorem 3.1 has been proved in [4] and [20], and in [13] in the case that the spaces X_i are pairwise homeomorphic.

3.2. *A Question.* The following question, raised in [4], apparently remains unsolved. We emphasize that, as with Question 2.3 above, a solution is desired in ZFC (without the assumption of special set-theoretic axioms).

Is the cardinal number 2^{2^ω} optimal in Theorem 3.1? Is there a family $\{X_i : i \in I\}$ of spaces, with $|I| = 2^{2^\omega}$, such that X_I is not countably compact but X_J is countably compact for all $J \subset I$ such that $\emptyset \neq J \neq I$? Is there a space X such that $X^{2^{2^\omega}}$ is not countably compact but X^α is countably compact for all cardinals $\alpha < 2^{2^\omega}$?

Reference is made in [5] to published and forthcoming works of van Douwen, of Juhász, Nagy and Weiss, of Kunen, of Rajagopalan, of Rajagopalan and Woods, and of Vaughan, which answer portions of these questions under a variety of set-theoretic assumptions known to be consistent with ZFC.

§4. *On Glazer's Proof of Hindman's Theorem.*

If S is a set we write

$$[S]^{\omega} = \{A \subset S : |A| = \omega\} \text{ and}$$

$$[S]^{<\omega} = \{A \subset S : |A| < \omega\}.$$

We denote the set of positive integers by $\mathbf{N}$, and for $F = \{k_n : n < m\} \in [\mathbf{N}]^{<\omega}$ we set $\Sigma F = \Sigma\{k_n : n < m\}$.

It was conjectured by Graham and Rothschild [16] that if $\mathbf{N} = A_0 \cup A_1$ then there are $k \in \{0,1\}$ and $B \in [A_k]^{\omega}$ such that $\Sigma F \in A_k$ for all $F \in [B]^{<\omega}$. The following statement, due to Hindman [17], establishes (a statement formally stronger than) the Graham-Rothschild conjecture. Hindman's proof [17] makes no use of ultrafilters, and will not concern us here. The proof we shall discuss is due to Glazer.

4.1. *Theorem.* If $n < \omega$ and $\mathbf{N} = \bigcup_{k<n} A_k$, then there are $k < n$ and $B \in [A_k]^{\omega}$ such that $\Sigma F \in A_k$ for all $F \in [B]^{<\omega}$.

Remarks Concerning the Proof. Define

$$\mathcal{A} = \{A \subset \mathbf{N} : \text{there is } B \in [A]^{\omega} \text{ such that } \Sigma F \in A \text{ for all } F \in [B]^{<\omega}\}.$$

It is enough to show that there is $p \in \beta(\mathbf{N})$ such that $p \subset \mathcal{A}$. To this end, for $A \subset \mathbf{N}$ and $n \in \mathbf{N}$ define

$$A - n = \{k \in \mathbf{N} : k + n \in A\}$$

and define an operation $\dot{+}$ on $\beta(\mathbf{N}) \times \beta(\mathbf{N})$ by

$$p \dot{+} q = \{A \subset \mathbf{N} : \{n \in \mathbf{N} : A - n \in p\} \in q\}.$$

It is easy to show that $\dot{+}$ is an associative function into $\beta(\mathbf{N})$ and that the function $q \to p \dot{+} q$ is, for each $p \in \beta(\mathbf{N})$, continuous as a function of q. It follows from Zorn's Lemma that there is $p \in \beta(\mathbf{N})$ such that $p \dot{+} p = p$; since $\dot{+}$ extends the usual addition function of $\mathbf{N}$, and since there is no $n \in \mathbf{N}$ such that $n + n = n$, we have $p \in \beta(\mathbf{N}) \setminus \mathbf{N}$.

We outline the proof that $p \subset A$. For $A \in p$ define

$$A^* = \{k \in \mathbb{N} : A - k \in p\},$$

and now fix $A = A_0 \in p$. Choose $k_0 \in A_0^* \cap A_0$ and recursively define $A_{n+1} = (A_n - k_n) \cap A_n$ and choose $k_{n+1} \in A_{n+1}^* \cap A_{n+1}$ so that $k_{n+1} > k_n$. Finally, define $B = \{k_n : n < \omega\}$. It is easily shown that $\Sigma F \in A_0 = A$ for all $F \in [B]^{<\omega}$.

Additional details of this proof are available in [14] and [5].

It is appropriate to note that it was F. Galvin who first raised the question whether there is $p \in \beta(\mathbb{N}) \setminus \mathbb{N}$ such that $\{n \in \mathbb{N} : A - n \in p\} \in p$ whenever $A \in p$, and who pointed out that an affirmative response would serve to establish the Graham-Rothschild conjecture; it was Glazer who defined the function $\dot{+}$ and showed the existence of $p \in \beta(\mathbb{N}) \setminus \mathbb{N}$ such that $p = p \dot{+} p$.

It has been pointed out to me by I. Prodanov and others that a proof which has been available for some years of Ramsey's theorem $\omega \rightarrow (\omega)^2_n$ is reminiscent of the proof just given. Indeed let $p \in U(\omega)$ and suppose that $[\omega]^2 = \bigcup_{k<n} A_k$. For $i < \omega$ define $A_k(i) = \{m < \omega : \{m,i\} \in A_k\}$ and set $B_k = \{i : A_k(i) \in p\}$. Since $\bigcup_{k<n} B_k = \omega$ there is $B_{\bar{k}} \in p$. Choose $k_0 \in B_{\bar{k}}$ and recursively choose $k_{m+1} \in B_{\bar{k}} \cap \bigcap_{j \leq m} A_{\bar{k}}(k_j)$ so that $k_{m+1} > k_m$. Then $B = \{k_m : m < \omega\}$ satisfies $B \in [\omega]^\omega$ and $[B]^2 \subset A_{\bar{k}}$. This is essentially the proof of the relation $\omega \rightarrow (\omega)^2_n$ given, for example, by Chang and Keisler [3] (Theorem 3.3.7) and by Jech [18] (Problem 7.5.1).

4.2. *Hindman's Theorem in* ZF. A mathematician reading Glazer's proof of Hindman's theorem may be dissatisfied because the proof appeals to the Axiom of Choice while the result itself "looks as if" it should be provable without appeal to that axiom. The same objection may be lodged against the original proof of Hindman [17] and the proof given subsequently by Baumgartner [1].

Responding to an inquiry whether Hindman's Theorem is a theorem in ZF, Professor Baumgartner has communicated the following information (letter of September, 1976); this is recorded here with his kind permission.

Hindman's Theorem is a Π^1_2 assertion, *i.e.*, it can be put in the form $(\forall X \in P(\omega))(\exists Y \in P(\omega))\phi(X,Y)$, where ϕ contains only first-order quantifiers ranging over the natural numbers. The following theorem, due to Shoenfield, asserts that Π^1_2 sentences are "absolute".

Theorem. Let M and N be transitive models of ZF (N may be the universe of set theory) such that $M \subset N$ and every countable ordinal of N lies in M. Then if ϕ is any Π^1_2-sentence, ϕ is true in M if and only if ϕ is true in N.

The standard reference is [21] (pp. 132-139), though the result given there is weaker than the theorem above.

To verify Hindman's Theorem it is enough to show that if ϕ is Π^1_2 and ZFC $\vdash \phi$, then ZF $\vdash \phi$. Suppose that ϕ is $(\forall X)(\exists Y)\psi(X,Y)$. Fix X and consider $L[X]$, the class of all sets constructible from X. It is well-known that $L[X] \models \phi$. Hence $\exists Y \in L[X]$ such that $L[X] \models \phi(X,Y)$. But since $\psi(X,Y)$ involves only first-order quantification over the integers, and since the integers are the same in $L[X]$ as in the "real world", it follows that $\psi(X,Y)$ is really true. Thus we have shown that $(\forall X)(\exists Y)\psi(X,Y)$, *i.e.*, that ϕ is true. Since this argument will work in any model of ZF, it follows that ZF $\vdash \phi$.

LIST OF REFERENCES

[1] James Baumgartner, *A short proof of Hindman's theorem,* J. Combinatorial Theory A 17 (1974), 384-386.

[2] Allen R. Bernstein, *A new kind of compactness for topological spaces,* Fundamenta Math. 66 (1970), 185-193.

[3] C. C. Chang and H. J. Keisler, Studies in Model Theory. Studies in Logic and the Foundations of Mathematics Vol. 73. North-Holland Publ. Co., Amsterdam-London, 1973.

[4] W. W. Comfort, Review of [13], Mathematical Reviews 52 #1633 (1976), 227-228.

[5] W. W. Comfort, *Ultrafilters: some old and some new results*, Bull. Amer. Math. Soc. (1977), (to appear).

[6] W. W. Comfort and S. Negrepontis, The Theory of Ultrafilters, Grundlehren der math. Wissenschaften vol. 211. Springer-Verlag, Berlin-Heidelberg-New York, 1974.

[7] B. A. Efimov, *On the embedding of extremally disconnected spaces into bicompacta*, In: General Topology and Its Relations to Modern Analysis and Algebra III. Proc. Third Prague Topological Symposium, Prague, 1971, edited by J. Novák, pp. 103-107. Academia Prague, Prague, 1972.

[8] Zdeněk Frolík, *Types of ultrafilters on countable sets*, In: General Topology and its Relations to Modern Analysis and Algebra II, Proc. Second Prague Topological Symposium, Prague, 1966, edited by J. Novák, pp. 142-143. Academia Prague, Prague, 1967.

[9] Zdeněk Frolík, *Homogeneity problems for extremally disconnected spaces*, Commentationes Math. Univ. Carolinae 8 (1967), 757-763.

[10] Z. Frolík, *Maps of extremally disconnected spaces, theory of types, and applications*, In: General Topology and its Relations to Modern Analysis and Algebra, Proc. (1968) Kanpur Topological Conference, pp. 131-142. Academia Prague, Prague, 1971.

[11] L. Gillman and M. Henriksen, *Rings of continuous functions in which every finitely generated ideal is principal*, Trans. Amer. Math. Soc. 82 (1956), 366-391.

[12] Leonard Gillman and Meyer Jerison, Rings of Continuous Functions, D. Van Nostrand Co., Inc., Princeton, 1960.

[13] John Ginsburg and Victor Saks, *Some applications of ultrafilters in topology*, Pacific J. Math. 57 (1975), 403-418.

[14] Steven Glazer, Manuscript, (to appear).

[15] Irving Glicksberg, *Stone-Čech compactifications of products*, Trans. Amer. Math. Soc. 90 (1959), 369-382.

[16] R. L. Graham and B. L. Rothschild, *Ramsey's theorem for n-parameter sets*, Trans. Amer. Math. Soc. 159 (1971), 257-292.

[17] Neil Hindman, *Finite sums from sequences within cells of a partition of* $\mathbb{N}$, J. Combinatorial Theory (A) 17 (1974), 1-11.

[18] Thomas J. Jech, The Axiom of Choice. Studies in Logic and the Foundations of Mathematics vol. 75. North-Holland Publ. Co., Amsterdam-London, 1973.

[19] Kenneth Kunen, *Ultrafilters and independent sets*, Trans. Amer. Math. Soc. 172 (1972), 299-306.

[20] Victor Saks, *Ultrafilter invariants in topological spaces*, Manuscript, (to appear).

[21] J. R. Shoenfield, *The problem of predicativity*, In: Essays on the Foundations of Mathematics, pp. 132-139. Magnes Press, Hebrew University, Jerusalem, 1966.

SOME PROBLEMS CONCERNING C(X)

Á. Császár
Loránd Eötvös University
H-1088 Budapest, Muzeum körut 6-8, Hungary

0. In this paper, some existence problems are investigated concerning function classes having the form $C(X)$, $C(Y)|X = \{f|X: f \in C(Y)\}$ or $C(Y) \circ p = \{f \circ p: f \in C(Y)\}$ where $p: X \to Y$ is a given map. The following questions are treated. Given a set X and a class Φ of real-valued functions defined on X, is there a topology on X such that $\Phi = C(X)$ (or on a set Y such that $\Phi = C(Y) \circ p$ for some surjective map $p: X \to Y$, or on a set $Y \supset X$ such that $\Phi = C(Y)|X$)? Given a topological space X and a Φ as above, is there a space $Y \supset X$ containing X as a subspace such that $\Phi = C(Y)|X$? Given a ring A, is there a topological space X (or a topological space Y and a subspace $X \subset Y$) such that A be isomorphic with $C(X)$ (or $C(Y)|X$)?

In answering (or partially answering) questions of this kind, we will use some types of function classes defined in [5] and [14].

1. Let X be an arbitrary non-empty set. Φ is said to be a function class on X if each $f \in \Phi$ is a real-valued function $f: X \to R$ defined on X. If $S \neq \emptyset$ and $p: S \to X$, we denote $\Phi \circ p = \{f \circ p: f \in \Phi\}$; for $\emptyset \neq T \subset X$, define $\Phi | T = \{f|T: f \in \Phi\}$. A function ring or a function lattice on X is a function class on X that is a ring or a lattice, respectively, under pointwise defined operations $+, \cdot, \vee, \wedge$, and that contains all constant functions. An affine lattice on X is a function lattice Φ on X such that $f \in \Phi$, $c \in R$ implies $f + c \in \Phi$, $cf \in \Phi$.

If X is a topological space with topology τ, $C(X)$ or $C(\tau)$ denotes the class of all continuous real-valued functions on X. Clearly $C(X)$ is a function ring and an affine lattice on X, and the same holds for $C(Y)|X$ or $C(Y) \circ p$ with $p: X \to Y$.

A function class Φ on X is said to be strongly composition-closed (scc), or composition-closed (cc) ([5], pp. 143 and 146), or weakly composition-closed (wcc) ([14], p. 114) if the following is true: given a family $\{f_i: i \in I\} \subset \Phi$, consider the map

(1.1) $\qquad h: X \to R^I, \quad h(x) = (f_i(x)) \quad \text{for} \quad x \in X,$

and a function $k \in C(h(X))$, or $k \in C(\overline{h(X)})$, or $k \in C(R^I)$ respectively; then $k \circ h \in \Phi$. Here R^I is equipped with the product topology, $\overline{Z}$ denotes the closure with respect to this topology, $h(X)$ and $\overline{h(X)}$ are considered as subspaces of R^I.

If the above conditions are restricted by the assumption that the index set I is countable, then Φ is said to be <u>countably strongly composition-closed</u> (cscc), <u>countably composition-closed</u> (ccc), <u>countably weakly composition-closed</u> (cwcc) respectively. Similarly, if I is supposed to be finite, Φ is said to be <u>finitely strongly composition-closed</u> (fscc), <u>finitely composition-closed</u> (fcc), or <u>finitely weakly composition-closed</u> (fwcc). For these classes, the following implications hold ([5], p. 147):

$$\begin{array}{ccccc} \text{scc} & \Rightarrow & \text{cc} & \Rightarrow & \text{wcc} \\ \Downarrow & & \Downarrow & & \Updownarrow \\ \text{cscc} & \Rightarrow & \text{ccc} & \Leftrightarrow & \text{cwcc} \\ \Downarrow & & \Downarrow & & \Downarrow \\ \text{fscc} & \Rightarrow & \text{fcc} & \Leftrightarrow & \text{fwcc} \end{array}$$

Φ is said to be <u>inversion-closed</u> if $f \in \Phi$, $f(x) \neq 0$ for $x \in X$ implies $1/f \in \Phi$.

If $f: X \to R$, let us denote $Z(f) = \{x \in X: f(x) = 0\}$; $Z(f)$ is the <u>zero-set of the function</u> f. If Φ is a function class on X, denote $Z(\Phi) = \{Z(f): f \in \Phi\}$. If X is a topological space with topology τ, we write $Z(X) = Z(C(X))$ or $Z(\tau) = Z(C(\tau))$; the elements of $Z(X)$ are the <u>zero-sets of the space</u> X. For $f: X \to R$, $c \in R$, put $X(f \geqq c) = \{x \in X: f(x) \geqq c\}$, $X(f \leqq c) = \{x \in X: f(x) \leqq c\}$. The sets $X(f \geqq c)$ and $X(f \leqq c)$ are called <u>level sets</u> (or <u>Lebesgue sets</u>) of f.

A function class Φ on X is said to be <u>complete</u> (or <u>uniformly closed</u>) if $f_n \in \Phi$ implies $f \in \Phi$ whenever $f_n \to f$ uniformly on

<u>2</u>. We start by examining the first type of questions formulated in the introduction. Let Φ be a function class on a set X. Under what conditions does a topology exist on the set X such that $\Phi = C(X)$? This question is answered by the following

<u>Theorem 1</u>. If X is a topological space and $\Phi = C(X)$, then

(a) Φ is a function ring,

(a') Φ is an affine lattice,

(b) $f = \{\sup g_i: i \in I\} = \inf\{h_j: j \in J\}$, $g_i, h_j \in \Phi$, $I \neq \emptyset$, $J \neq$ implies $f \in \Phi$,

(c) $(f \vee (-c)) \wedge c \in \Phi$ for $c \in R$ implies $f \in \Phi$, (c') Φ is inversion-closed.

Conversely if Φ is a function class on X satisfying either (a), (b), (c), or (a), (b), (c'), or (a'), (b), (c), then there is a topology on X such that $\Phi = C(X)$.

Proof. The necessity of (a) to (c') can be easily checked (see [4], pp. 184-185). If Φ satisfies (b) and either (a) or (a'), then the same holds for the class Φ^* composed of all bounded functions in Φ ; hence, by [4], Theorem 8, $\Phi^* = C^*(\tau)$ for some topology τ on X. In the case of (a') we easily get that $f \in \Phi$ implies $f_c = (f \vee (-c)) \wedge c \in \Phi^* = C^*(\tau) \subset C(\tau)$, hence $f \in C(\tau)$; conversely $f \in C(\tau)$ implies $f_c \in C^*(\tau) = \Phi^* \subset \Phi$ and, by (c), $f \in \Phi$. In the case of (a), the function $h: R \to R$, $h(u) = |u|$, can be represented as $h(u) = u \vee (-u)$ and also as the infimum of a family of quadratic polynomials; therefore (b) implies that $|f| \in \Phi$ whenever $f \in \Phi$, i.e. that Φ is a lattice. Hence the case of (a), (b), (c) has been reduced to that one of (a'), (b), (c). Finally, in the case of (a), (b), (c'), $f \in \Phi$ implies $g = f/(1 + |f|) \in \Phi^* = C^*(\tau)$, hence $f = g/(1 - |g|) \in C(\tau)$. Conversely $f \in C(\tau)$ implies $g \in C^*(\tau) = \Phi^* \subset \Phi$ and $f \in \Phi$.

Remark 1. By [4], p. 184, (b) implies that Φ is complete; hence (a), (b), (c') $\Rightarrow$ (c) is a consequence of [17], Corollary 3.7, and we get an alternative proof for the case (a), (b), (c').

Another answer to the above question is contained in

Theorem 2. If Φ is a function class on X, the following statements are equivalent:

(a) There are a topological space Y and a surjective map $p: X \to Y$ such that $\Phi = C(Y) \circ p$,

(b) Φ is strongly composition-closed,

(c) There is a topology τ on X such that $\Phi = C(\tau)$.

Proof. (a) $\Rightarrow$ (b): Suppose $f_i \in \Phi$, $f_i = g_i \circ p$, $g_i \in C(Y)$, $i \in I$. Define $h: X \to R^I$ and $g: Y \to R^I$ by $h(x) = (f_i(x))$, $g(y) = (g_i(y))$. Then $h = g \circ p$ and g is continuous, hence if $k \in C(h(X)) = C(g(Y))$, we have $k \circ g \in C(Y)$ and $k \circ h \in C(Y) \circ p$.

(b) $\Rightarrow$ (c) is contained in [5], (2.6), and (c) $\Rightarrow$ (a) is obvious.

A similar argument furnishes, using [14], p. 114 for (b) $\Rightarrow$ (c):

Theorem 3. For a function class on X, the following statements are equivalent:

(a) There are a topological space Y and a map $p: X \to Y$ such that $\Phi = C(Y) \circ p$,

(b) Φ is weakly composition-closed,

(c) There are a set $Y \supset X$ and a topology on Y such that $\Phi = C(Y) | X$.

A modification of this theorem yields:

<u>Theorem 4</u>. For a function class Φ, the following statements are equivalent:

(a) There are a topological space Y and a map $p: X \to Y$ such that $p(X)$ is dense in Y and $\Phi = C(Y) \circ p$,

(b) Φ is composition-closed,

(c) There are a set $Y \supset X$ and a topology on Y such that X is dense in Y and $\Phi = C(Y) | X$.

<u>Proof</u>. (a) $\Rightarrow$ (b): Similarly as in the proof of Theorem 2, by $g(Y) = g(\overline{p(X)}) \subset \overline{g(p(X))} = \overline{h(X)}$.

(b) $\Rightarrow$ (c): Suppose that Φ is cc, set $\Phi = \{f_i : i \in I\}$ with some index set I, and consider h as in (1.1). Choose a set $Y \supset X$ such that there exists a bijection $h': Y - X \to Z - h(X)$ where $Z = \overline{h(X)}$, and define $q: Y \to Z$ by $q(x) = h(x)$ for $x \in X$, $q(x) = h'(x)$ for $x \in Y - X$. Equip Y with the inverse image by q of the topology of Z (i. e. $G \subset Y$ is open iff $G = q^{-1}(H)$ for some $H \subset Z$ open in Z). Now if $f \in \Phi$, say $f = f_i$, $i \in I$, then $f_i = p_i \circ q | X$ where p_i denotes the projection of R^I onto its i-th factor, hence $p_i \circ q \in C(Y)$. Conversely if $g \in C(Y)$, then $x, y \in Y$, $g(x) \neq g(y)$ implies that x and y have disjoint neighbourhoods in Y so that $q(x) \neq q(y)$; therefore $g = k \circ q$ with $k \in C(Z)$ and $g|X = k \circ h \in \Phi$ by the cc property of Φ. Finally X is dense in Y since $h(X) = q(X)$ is dense in $Z = \overline{h(X)}$.

(c) $\Rightarrow$ (a): Obvious.

<u>Remark 2</u>. The proofs of (b) $\Rightarrow$ (c) in Theorems 2 and 3 are quite similar to that one in Theorem 4, only one has to choose $Z = h(X)$ or $Z = R^I$, respectively.

3. Now let X be a given topological space, Φ a function class on X, and let us look for conditions on Φ involving $\Phi = C(X)$ or $\Phi = C(Y) | X$ with a suitable topological space Y containing X as a subspace. Theorems 2, 3, 4 furnish results of this kind provided the spaces in question are completely regular (not necessarily Haus-

dorff).

Theorem 5. Let X be a completely regular space and Φ a function class on X. $\Phi = C(X)$ iff Φ is strongly composition-closed and $Z(\Phi)$ is a closed base in X (i. e. a system of closed sets such that each closed set is the intersection of some elements of this system).

Proof. The necessity follows from Theorem 2 and from the fact that a space is completely regular iff $Z(X)$ is a closed base ([7], p. 38). Conversely if Φ is scc, then, by Theorem 2, it coincides with $C(\tau)$ for a topology τ obtained as the inverse image topology of a subspace of R^I (see Remark 2). Consequently τ is completely regular, thus $Z(\Phi)$ is a closed base for τ, and τ is identical with the given topology of X since both topologies admit the same closed base.

Theorem 6. Let X be a topological space and Φ a function class on X. There is a completely regular space $Y \supset X$ containing X as a subspace and satisfying $\Phi = C(Y)|X$ iff Φ is weakly composition-closed and $Z(\Phi)$ is a closed base in X.

Proof. If $\Phi = C(Y)|X$ and Y is completely regular, then Φ is wcc by Theorem 3 and $Z(\Phi)$, being the trace on X of $Z(Y)$, is a closed base in X. Conversely if Φ is wcc, then $\Phi = C(Y)|X$ by Theorem 3 for a suitable set $Y \supset X$ and a completely regular topology τ on Y (by Remark 2). Hence $Z(Y)$ is a closed base for τ and the same holds for $Z(\Phi)$ and the topology induced by τ on X; consequently the latter coincides with the given topology of X.

A similar argument, using Theorem 4 instead of Theorem 3, furnishes

Theorem 7. Let X be a topological space and Φ a function class on X. There is a completely regular space $Y \supset X$ containing X as a dense subspace and satisfying $\Phi = C(Y)|X$ iff Φ is composition-closed and $Z(\Phi)$ is a closed base in X.

Theorem 8. Let X be a T_0-space and Φ a function class on X. There is a completely regular Hausdorff space $Y \supset X$ containing X as a (dense) subspace iff Φ is weakly composition-closed (composition-closed) and $Z(\Phi)$ is a closed base in X.

Proof. By the proof of Theorem 6 (or 7) the topology on Y is now Hausdorff since the map $q: Y \rightarrow Z$ in the proof of Theorem 4 is injective.

Remark 3. The same argument shows that, under the hypotheses of Theorem 8, the space Y can be supposed to be realcompact (see [7], 8.2, 8.10, 8.11; cf. [5], (3.5)).

If the topology given on X is not completely regular, then the condition that $Z(\Phi)$ be a closed base in X is no more necessary for having $\Phi = C(X)$ or $\Phi = C(Y)|X$. However, Theorem 5 can be modified in a suitable way in order to embrace the case of an arbitrary topological space X:

Theorem 9. Let X be a topological space and Φ a function clas on X. $\Phi = C(X)$ iff Φ is (countably) strongly composition-closed and $Z(\Phi) = Z(X)$.

Proof. If $\Phi = C(X)$, then Φ is scc (and a fortiori cscc) by Theorem 2 and, of course, $Z(\Phi) = Z(X)$. Conversely if Φ is scc, then $\Phi = C(\tau)$ for a topology τ on X, and $C(\tau)$ is composed of all functions $f: X \rightarrow R$ the level sets of which belong to $Z(\tau) = Z(\Phi)$. Similarly $C(X)$ is the class of all functions with level sets in $Z(X)$, hence $Z(\Phi) = Z(X)$ implies $C(\tau) = C(X)$ and $\Phi = C(X)$.

If Φ is cscc, then, by Theorem 10 below, Φ is still composed of all functions the level sets of which belong to $Z(\Phi)$ so that $Z(\Phi) = Z(X)$ implies again $\Phi = C(X)$.

Theorem 10. ([6], Theorem 16). Let Φ be a function class on X Then the following statements are equivalent:

(a) Φ is countably strongly composition-closed,

(b) Φ is finitely strongly composition-closed and complete,

(c) Φ is finitely composition-closed, inversion-closed, and complete,

(d) Φ is a complete and inversion-closed function ring,

(e) $Z(\Phi)$ is a δ-lattice in X (i. e. $\emptyset, X \in Z(\Phi)$, $Z_1, Z_2 \in Z(\Phi)$ implies $Z_1 \cup Z_2 \in Z(\Phi)$, $Z_i \in Z(\Phi)$ for $i = 1, 2, \ldots$ impli $\bigcap\{Z_i: i = 1, 2, \ldots\} \in Z(\Phi)$), and Φ is composed of all functions the level sets of which belong to $Z(\Phi)$,

(f) There is a δ-lattice in X so that Φ coincides with the class of all functions the level sets of which belong to this δ-lattice.

Remark 4. (a) $\Rightarrow$ (b) follows from [13], Theorem 2.1, (b) $\Rightarrow$ (c) $\Rightarrow$ (d) is obvious, (d) $\Leftrightarrow$ (e) $\Leftrightarrow$ (f) is essentially contained in [11], pp. 236 and 241 (cf. also [17], Theorem B' and [10], Lemma 6.3), finally (d) $\Rightarrow$ (a) is [16], Theorem 4.9.

An interesting consequence of Theorem 10 is the fact that the cscc classes can be characterized as fscc classes having some approximation property (namely completeness), and even the fscc property can be reduced to the condition of being closed under a finite number of suitable operations. Theorems 1 and 2 yield a similar characterization of scc classes.

Problem 1. Is it possible to characterize the cc and wcc classes by the property of being fcc and by suitable approximation properties?

In this direction it is worth-wile to mention that every wcc class is complete ([13], Theorem 2.1).

Problem 2. Let X be an arbitrary topological space and Φ a function class on X. Give necessary and sufficient conditions in order to have $\Phi = C(Y)|X$ for a suitable topological space $Y \supset X$ containing X as a (dense) subspace.

Notice that, in Theorems 6 and 7, complete regularity of Y was essential to assure the necessity of the given conditions.

4. Let now A be an arbitrary ring. We shall study the question under what conditions is A isomorphic with a ring of the form $C(X)$ or $C(Y)|X$. First we examine the (purely algebraic) question when is A isomorphic with a function ring. The following theorem is an easy consequence of well-known results on subdirect sums of rings (see e. g. [15], Theorem 2, or [18]), and is essentially known ([1], [2], [12], [14]):

Theorem 11. For a ring A, there is a function ring isomorphic with A iff the conditions (a), (b), and one of the conditions (c_i) $(i = 1, 2, 3)$ below are fulfilled.

(a) A contains a unit element $e \neq 0$,

(b) There is a homomorphism $\chi : R \rightarrow A$ such that $e \in \chi(R)$,

(c_1) For every (two-sided) ideal $I \subset A$, $I \neq \{0\}$, there exists a homomorphism $\varphi : I \rightarrow R$ such that $1 \in \varphi(I)$,

(c_2) For every ideal $I \subset A$, $I \neq \{0\}$, there is a homomorphism $s: A \to R$ such that $s(I) \neq \{0\}$,

(c_3) For every $a \in A$, $a \neq 0$, there is a homomorphism $s: A \to R$ such that $s(a) \neq 0$.

Under these hypotheses, a function ring isomorphic with A can be constructed over the set S of all homomorphisms $s: A \to R$ such that $s(A) \neq \{0\}$ by assigning to $a \in A$ the function $f: S \to R$ defined by $f(s) = s(a)$.

Proof. Assume that Φ is a function ring on X. Then (a), (b), (c_1) hold for Φ instead of A, consequently for any ring A isomorphic with Φ. In fact, the constant 1 is the unit of Φ, the homomorphism χ assigning the constant c to $c \in R$ satisfies (b), and if I is an ideal in Φ, $g \in I$, $g \neq 0$, then there is $x \in X$ with $g(x) \neq 0$, and $\varphi(f) = f(x)$ $(f \in I)$ defines a homomorphism $\varphi: I \to R$ satisfying (c_1) since $\varphi(g/c) = 1$ for $c = g(x)$.

For a ring A, (c_1) $\Rightarrow$ (c_2). For, if I is an ideal in A and $\varphi: I \to R$ is a homomorphism, $i \in I$, $\varphi(i) = 1$, then $s(a) = \varphi(ia)$ defines a homomorphism $s: A \to R$ for which $s(i) = 1$.

(c_2) $\Rightarrow$ (c_3) for any ring A. In fact, if (c_3) were false, then $a \neq 0$ would be contained in the intersection J of the kernels of all homomorphisms $s: A \to R$ with $s(A) \neq \{0\}$ (the existence of at least one s of this kind follows from (c_2) for $I = A \neq \{0\}$). By (c_2), there would be a homomorphism $s: A \to R$ such that $s(J) \neq \{0\}$ which contradicts to $J \subset \text{Ker } s$.

Finally if A satisfies (a), (b), (c_3), then denote by S the set of all homomorphisms $s: A \to R$ such that $s(A) \neq \{0\}$, and define $f_a(s) = s(a)$ for $a \in A$, $s \in S$. Then $S \neq \emptyset$ by (a) and (c_3), and clearly $f_{a+b} = f_a + f_b$, $f_{ab} = f_a f_b$ so that $\Phi = \{f_a: a \in A\}$ is a ring under pointwise addition and multiplication, and $\psi(a) = f_a$ defines an epimorphism $\psi: A \to \Phi$. Moreover, Φ is a function ring (i. e. it contains all constant functions). In fact, if $s \in S$, then $s \circ \chi: R \to R$ is a homomorphism such that $s(\chi(R)) \neq \{0\}$ since $e \in \chi(R)$ by (b) and $s(e) = 0$ would imply $s(a) = s(ae) = s(a)s(e) = 0$ for each $a \in A$. Therefore $s(\chi(c)) = c$ for $c \in R$ ([7], 0.22), i. e. $\psi(\chi(c)) \in \Phi$ is the constant function c. Finally ψ is a monomorphism by (c_3).

Theorem 11, combined with Theorems 2 and 3, furnishes characterizations of those rings that are isomorphic with rings of the form $C(X)$ or $C(Y)|X$. The first one of these problems was investigated by many authors (see e. g. [12],[1],[2],[9]). We shall need two lemma that are slight modifications of [7], 3.9 and 8.8, and [7], 0.23 and 8.3, respectively.

Lemma 1. Let Y be a topological space and $X \subset Y$. Then there exist a realcompact Tychonoff space Y' and a closed subspace $X' \subset Y'$ such that the rings $C(Y)$ and $C(Y)|X$ are isomorphic with $C(Y')$ and $C(Y')|X'$ respectively.

Proof. Let Y_1 be the set Y equipped with the topology for which $Z(Y)$ is a closed base, and $X_1 = X$. Then $C(Y_1) = C(Y)$, and $C(Y_1)|X_1 = C(Y)|X$. Moreover, Y_1 is completely regular.

Now let Y_2 be the set of all equivalence classes belonging to the equivalence relation introduced in Y_1 by setting $x \sim y$ iff $f(x) = f(y)$ for every $f \in C(Y_1)$. Denote by $q(x)$, for $x \in Y_1$, the equivalence class containing x. Then equip Y_2 with the quotient topology corresponding to $q\colon Y_1 \to Y_2$, and define $X_2 = q(X_1)$. By assigning $f \circ q$ $(f \circ q|X_1)$ to $f \in C(Y_2)$ $(f|X_2 \in C(Y_2)|X_2)$, we obtain an isomorphism from $C(Y_2)$ onto $C(Y_1)$ (from $C(Y_2)|X_2$ onto $C(Y_1)|X_1$). Moreover, Y_2 is a Tychonoff space.

Finally let Y' denote the Hewitt realcompactification of Y_2 and X' the closure of X_2 in Y'. Then by assigning $f|Y_2$ to $f \in C(Y')$ $(f|X_2$ to $f|X' \in C(Y')|X')$, we get an isomorphism from $C(Y')$ onto $C(Y_2)$ (from $C(Y')|X'$ onto $C(Y_2)|X_2$).

Lemma 2. Let Y be a realcompact Tychonoff space and $X \subset Y$ be closed. Then the ring homomorphisms $s\colon C(Y)|X \to R$ such that $s(C(Y)|X) \neq \{0\}$ correspond in a one-to-one manner to the points of X, the homomorphism s corresponding to $x \in X$ being defined by $s(f|X) = f(x)$ for $f \in C(Y)$.

Proof. If $f \in C(Y)$, $s(f|X) = c \neq 0$, then $s(rf|X) = rc$ for $r \in R$ so that s is an epimorphism onto R. Denote by $k\colon C(Y) \to C(Y)|X$ the epimorphism defined by $k(f) = f|X$. Then $s \circ k$ is an epimorphism from $C(Y)$ onto R, hence $\mathrm{Ker}(s \circ k)$ is a real maximal ideal of $C(Y)$, and $s(k(f)) = f(x)$ for some $x \in Y$ depending on s and every $f \in C(Y)$. Clearly $x \in X$ since otherwise there would be an $f \in C(Y)$ vanishing on X but satisfying $f(x) = 1$. Thus $s(f|X) = f(x)$ for some $x \in X$ and every $f \in C(Y)$. Conversely, this equality defines an epimorphism $s\colon C(Y)|X \to R$, and distinct points x_1 and x_2 generate distinct epimorphisms.

The desired characterization of the rings isomorphic to rings of the form $C(X)$ is now the following:

Theorem 12. A ring A is isomorphic with a ring of the form $C(X)$ iff A satisfies conditions (a), (b), (c_i) $(i = 1, 2, 3)$ of Theorem 11, and the following is true: whenever $a_i \in A$ $(i \in I)$ and S denotes the set of all homomorphisms $s: A \to R$ such that $s(A) \neq \{0\}$, further $h: S \to R^I$ is defined by $h(s) = (s(a_i))$, finally $k \in C(h(S))$, then there is an $a \in A$ such that $k(h(s)) = s(a)$ for every $s \in S$.

Proof. If these conditions are fulfilled, then, by Theorem 11, an isomorphism $\psi: A \to \Phi$ is given by $\psi(a) = f$ where f is a function defined on S such that $f(s) = s(a)$ for $s \in S$, and $\Phi = \psi(A)$. Now our hypotheses assure precisely that Φ is scc so that, by Theorem 2, $\Phi = C(\tau)$ for a suitable topology on S.

Conversely let $\omega: C(X) \to A$ be an isomorphism, X a topological space. By Lemma 1 we can suppose that X is realcompact. Then, by Theorem 11, A fulfils (a), (b), (c_i). Moreover, $\{s \circ \omega: s \in S\}$ is the set of all non-vanishing homomorphisms from $C(X)$ into R, hence, by Lemma 2, we obtain a one-to-one correspondence between the homomorphisms $s \circ \omega$ and the points $x \in X$ by putting $s(\omega(f)) = f(x)$. By Theorem 2, $C(X)$ is scc, and this yields the last condition for A.

A similar argument furnishes, by using Theorem 3 instead of Theorem 2, and taking into account that the wcc and cwcc properties coincide:

Theorem 13. A ring A is isomorphic with a ring of the form $C(Y)|X$ iff A satisfies conditions (a), (b), (c_i) $(i = 1, 2, \text{or } 3)$ of Theorem 11, and the following is true: whenver $a_i \in A$ $(i = 1, 2, 3, \ldots)$ and S is the same as in Theorem 11, further $h: S \to R^N$ is defined by $h(s) = (s(a_i))$, finally $k \in C(R^N)$, then there is an $a \in A$ such that $k(h(s)) = s(a)$ for every $s \in S$.

5. Finally we investigate the following problem. Given a function ring Φ on a set T, look for conditions assuring that Φ be isomorphic with a ring of the form $C(X)$ or $C(Y)|X$. For this purpose, we need a slight modification of [7], 10.6:

Lemma 3. Let Φ be a function ring on a set T and Z a realcompact Tychonoff space. If $\omega: C(Z) \to \Phi$ is an epimorphism, then there is a map $p: T \to Z$ such that $\omega(g) = g \circ p$ for $g \in C(Z)$. If ω is an isomorphism, then $p(T)$ is dense in Z.

Proof. For $t \in T$, $s(g) = \omega(g)(t)$ defines an epimorphism s:

$C(Z) \to R$. Hence there is a unique $z \in Z$ satisfying $\omega(g)(t) = g(z)$. Define $z = p(t)$; then clearly $\omega(g) = g \circ p$. If $p(T)$ is not dense in Z, then there are $g_1, g_2 \in C(Z)$, $g_1 \neq g_2$ with $g_1 \circ p = g_2 \circ p$ and ω cannot be an isomorphism.

Theorem 14. Let Φ be a function ring on a set T. There is a ring $C(X)$ isomorphic with Φ iff Φ is composition-closed.

Proof. If Φ is cc, then, by Theorem 4, there is a topological space X containing T as a dense subset and satisfying $\Phi = C(X)|T$. Then clearly $C(X)$ is isomorphic with Φ, an isomorphism being obtained by assigning $f|T$ to $f \in C(X)$.

Conversely suppose that there is an isomorphism $\omega\colon C(Z) \to \Phi$. By Lemma 1, we can suppose that Z is a realcompact Tychonoff space. Consider the map $p\colon T \to Z$ of Lemma 3, and choose a set $X \supset T$ such that there is a bijection $p'\colon X - T \to Z - p(T)$. Define $q\colon X \to Z$ by $q(x) = p(x)$ for $x \in T$, $q(x) = p'(x)$ for $x \in X - T$. Equip X with the inverse image by q of the topology of Z. Then T is dense in X, and the elements of $C(X)$ are precisely the functions $g \circ q$ where $g \in C(Z)$. Clearly $\omega(g)(t) = g(p(t))$ for $g \in C(Z)$, $t \in T$, hence $\omega(g) = g \circ q|T$ and $\Phi = C(X)|T$ so that Φ is cc by Theorem 4.

Theorem 15. A function ring Φ on a set T is isomorphic with a ring of the form $C(Y)|X$ iff Φ is weakly composition-closed.

Proof. If Φ is wcc, then Theorem 3 yields $\Phi = C(Y)|T$ for some topological space $Y \supset T$. Conversely, if $\chi\colon C(Z)|Y \to \Phi$ is an isomorphism, then, by Lemma 1, Z can be supposed to be realcompact and Tychonoff. Define $\varphi\colon C(Z) \to C(Z)|Y$ by $\varphi(g) = g|Y$ and apply Lemma 3 for the epimorphism $\omega = \chi \circ \varphi\colon C(Z) \to \Phi$. Define p, X, p', q and the topology on X as in the proof of Theorem 14. Then Φ is composed of the functions $\omega(g)$ and $C(X)$ of those $g \circ q$ for $g \in C(Z)$, moreover $\omega(g) = g \circ q|T$. Hence $\Phi = C(X)|T$, and Φ is wcc by Theorem 3.

Remark 5. Theorem 15 can be obtained with the help of some results on uniform spaces ([8], [3]).

Remark 6. The function ring composed of all polynomials on the real line is not wcc (in fact, it is not complete), hence this ring is not isomorphic with any ring $C(Y)|X$. On the other hand, if Φ is the first Baire class on R, then Φ is wcc (in fact, it is cscc by Theorem 10) without being cc ([5], (2.5)), hence Φ is of the form

$C(Y)|R$ for a suitable space $Y \supset R$ without being isomorphic with a ring $C(X)$. A function ring Φ on T that is cc without being scc (cf. [5], p. 147) is isomorphic with a ring $C(X)$ but fails to be of the form $C(T)$.

References

[1] F. W. Anderson: Approximation in systems of real-valued continuous functions. Trans. Amer. Math. Soc. 103 (1962), 249-271.

[2] F. W. Anderson and R. L. Blair: Characterizations of the algebra of all real-valued continuous functions on a completely regular space. Illinois J. Math. 3 (1959), 121-133.

[3] H. H. Corson and J. R. Isbell: Some properties of strong uniformities. Quart. J. Math. Oxford 11 (1960), 17-33.

[4] Á. Császár: On approximation theorems for uniform spaces. Acta Math. Acad. Sci. Hungar. 22 (1971), 177-186.

[5] Á. Császár: Function classes, compactifications, realcompactifications. Ann. Univ. Sci. Budapest. Eötvös Sect. Math. 17 (1974), 139-156.

[6] Á. Császár and M. Laczkovich: Discrete and equal convergence. Studia Sci. Math. Hungar. (in print).

[7] L. Gillman and M. Jerison: Rings of continuous functions. D. Van Nostrand Company, Princeton-Toronto-London-New York, 1960.

[8] A. W. Hager: Three classes of uniform spaces. Proc. Third Prague Topol. Symp. 1971, 159-164.

[9] A. W. Hager: Vector lattices of uniformly continuous functions and some categorical methods in uniform spaces. TOPO 72, Second Pittsburgh Intern. Conf. 1972, 172-187.

[10] A. W. Hager: Some nearly fine uniform spaces. Proc. London Math. Soc. 28 (1974), 517-546.

[11] F. Hausdorff: Mengenlehre. Walter de Gruyter and Co., Berlin-Leipzig, 1927.

[12] M. Henriksen and D. G. Johnson: On the structure of a class of archimedean lattice-ordered algebras. Fund. Math. 50 (1961), 73-94

[13] M. Henriksen, J. R. Isbell and D. G. Johnson: Residue class fields of lattice-ordered algebras. Fund. Math. 50 (1961), 107-117.

[14] J. R. Isbell: Algebras of uniformly continuous functions. Ann. of Math. 68 (1958), 96-125.

[15] N. H. McCoy: Subdirect sums of rings. Bull. Amer. Math. Soc. 53 (1947), 856-877.

[16] S. Mrówka: On some approximation theorems. Nieuw Arch. Wisk.

(3) 16 (1968), 94-111.

[17] S. Mrówka: Characterization of classes of functions by Lebesgue sets. Czechoslovak Math. J. 19 (94) (1969), 738-744.

[18] R. Wiegandt: Problème 196. Mat. Lapok 24 (1973), 156.

Generalized Shape Theory

by

Aristide Deleanu and Peter Hilton

1. Introduction

Since Borsuk [1] first introduced the concept of shape in his study of the homotopy theory of compacta many authors (see, for example,[5,6,7,10,11,13,14,15,17]) have contributed to the development of shape theory. However the theory has remained almost exclusively confined to a topological context, never very far removed from the setting in which it was originally cast by Borsuk; and, further, and arising from this restriction in the scope of the theory, the concept has, in the work cited, related to some category of topological spaces $\mathfrak{T}$ and a full subcategory $\mathfrak{P}$ of $\mathfrak{T}$. However, Holsztynski [16] observed, soon after Borsuk's invention of the concept, that shape could be formulated as an abstract limit, and was thus of more general applicability.

It is the principal purpose of this paper to free shape theory from its restricted scope. Thus we replace the full embedding of a topological category $\mathfrak{P}$ in a topological category $\mathfrak{T}$ by an arbitrary functor $K : \mathfrak{P} \to \mathfrak{T}$ from the arbitrary category $\mathfrak{P}$ to the arbitrary category $\mathfrak{T}$. In so doing we are very much inspired by the point of view adopted by LeVan in his thesis [11]. We then find that many of the categorical aspects of shape theory (we do not speak of the topological aspects) remain valid in this very general setting. Others require some restriction on the functor K, but a restriction far milder than that K should be a full embedding.

We refer to the contribution of Sibe Mardešić to these proceedings for the foundations of shape theory. If $K : \mathfrak{P} \to \mathfrak{T}$ is the embedding of the homotopy category of compact polyhedra (or compact ANR's) in the homotopy category of compact Hausdorff spaces, then, basing himself on the Mardešić-Segal interpretation of Borsuk shape [15], via approximating ANR-systems, LeVan [11] showed the following. First, of course if $f : X \to Y$ is a map[1] in $\mathfrak{T}$, then f induces, for all objects P of $\mathfrak{P}$, a function $f^P : \mathfrak{T}(Y,P) \to \mathfrak{T}(X,P)$, simply by composition. Moreover, the functions f^P enjoy the naturality condition that, if $u : P \to Q$ is a map in $\mathfrak{P}$, then the diagram

(1.1)
$$\begin{array}{ccc} \mathfrak{T}(Y,P) & \xrightarrow{f^P} & \mathfrak{T}(X,P) \\ \downarrow u_* & & \downarrow u_* \\ \mathfrak{T}(Y,Q) & \xrightarrow{f^Q} & \mathfrak{T}(X,Q) \end{array}$$

commutes: here u_* is also induced by composition. Then LeVan's fundamental result in [11] is that a <u>shape morphism</u> from X to Y is nothing but a family of functions f^P, $P \in |\mathfrak{P}|$, such that (1.1) commutes for all $u : P \to Q$ in $\mathfrak{P}$. It is this point of view which we now adopt. Thus our generalization consists of replacing the special functor K by an arbitrary functor K between arbitrary categories $\mathfrak{P}$ and $\mathfrak{T}$ and <u>defining</u> the shape category by the obvious generalization of LeVan's characterization. Explicitly, given a functor K from a category $\mathfrak{P}$ to a category $\mathfrak{T}$, we define $\mathfrak{S}$, the <u>shape category of K</u>, to be the category whose objects are those of $\mathfrak{T}$, with (reexpressing (1.1)).

[1] Notice that a <u>map</u> is here a homotopy class of continuous functions.

$$(1.2) \qquad \mathfrak{S}(X,Y) = \mathrm{Nat}(\mathfrak{T}(Y,K-),\mathfrak{T}(X,K-))$$

Moreover, it is plain from the discussion above that every morphism of $\mathfrak{T}$ induces a morphism of $\mathfrak{S}$, so that there is a canonical functor $T : \mathfrak{T} \to \mathfrak{S}$ which is the identity on objects. Precisely, we regard the pair $(\mathfrak{S},T)$ as the shape of K.

Plainly, this generalization substantially broadens the scope of shape theory. However, it also has another purpose, namely, to identify those parts of the existing theory which are "trivial" - and to prove them by appropriately "trivial" arguments - and thus to enable one to focus, in any particular concretization, on the deep aspects of the theory. We will exemplify this latter aspect in the next section. Then in Section 3 we will apply shape theory in new contexts, thus exhibiting connections between different mathematical theories which are perhaps not immediately evident. We emphasize that the role of our categorical formulations is as stated above, and not to prove known or unknown difficult theorems. By means of our generalization we establish connections and know, as a result, what questions to ask in various mathematical contexts; to the "non-trivial" aspects of the answers we do not claim that our approach contributes.

Details of some of our specific results are to be found in [2,3].

2. Universal properties of shape theory

The approach taken in [14] shows that, in the original context of shape theory, we have the result

$$\mathfrak{T}(X,P) = \mathfrak{S}(X,P);$$

that is, the shape morphisms from a compact space X to a compact ANR P are just the original maps from X to P in $\mathfrak{T}$. It turns out that this property requires a mild restriction on the functor K, which leads to a concept which proves relevant in many contexts.

Definition 2.1 The functor $K : \mathfrak{P} \to \mathfrak{T}$ is rich [2] if, given objects P,Q of $\mathfrak{P}$ and a morphism $f : KP \to KQ$ in $\mathfrak{T}$, there exists a path

$$P = V_o \xrightarrow{v_1} V_1 \xleftarrow{v_2} V_2 \longrightarrow \cdots \xrightarrow{v_{2k-1}} V_{2k-1} \xleftarrow{v_{2k}} V_{2k} = Q \quad \text{in} \quad \mathfrak{P},$$

such that each Kv_{2i} is invertible and

$$f = (Kv_{2k})^{-1} o Kv_{2k-1} o \cdots o (Kv_2)^{-1} o Kv_1 .$$

This definition is equivalent to the condition that, if we form the category of fractions $\mathfrak{P}[\Sigma^{-1}]$ with respect to the morphisms inverted by K, and if $\bar{K} : \mathfrak{P}[\Sigma^{-1}] \to \mathfrak{T}$ is induced by K, then $\bar{K}$ is full. An example of a rich functor which is not full is the direct limit functor from sequences of groups to groups.

Theorem 2.1 *If $K : \mathfrak{P} \to \mathfrak{T}$ is rich then* $T : \mathfrak{T}(X,KP) \to \mathfrak{S}(X,KP)$ *is bijective for all* X in $|\mathfrak{T}|$, P in $|\mathfrak{P}|$.

We would wish passage to the shape category to be idempotent. That is, if $K_1 = TK : \mathfrak{P} \to \mathfrak{S}$ we would wish $(\mathfrak{S},1)$ to be the shape of K_1. We find

Theorem 2.2 *If* $K : \mathfrak{P} \to \mathfrak{T}$ *is rich then shape is idempotent*

Indeed, as observed explicitly by A. Frei, the idempotence of shape follows from the conclusion of Theorem 2.1.

As a further example of a universal property of shape, consider the well-known result that, in the original restricted context of shape theory, Čech cohomology is shape-invariant. In our formulation, we say that a functor $G : \mathfrak{T} \to \mathfrak{C}$ is *shape-invariant* if it factors as $\bar{G}T$ with $\bar{G} : \mathfrak{S} \to \mathfrak{C}$. Plainly if G is shape-invariant then GX is equivalent to GY whenever X,Y have the same shape. Now Dold pointed out, in the appendix to [4], that Čech cohomology on the category of compact spaces is the right Kan extension [12] of ordinary (simplicial) cohomology on the category of compact polyhedra. Thus the shape-invariance of Čech cohomology is a special case of the following universal fact.

Theorem 2.3 Let $F : \mathfrak{P} \to \mathfrak{C}$ *be a functor and let* $\tilde{F} : \mathfrak{T} \to \mathfrak{C}$ *be the right Kan extension of* F *along* K. *Then* $\tilde{F}$ is *shape-invariant*.

In fact, there is a canonical factorization

$$\tilde{F} = \bar{F}T, \; \bar{F} : \mathfrak{S} \to \mathfrak{C},$$

and one easily proves

Theorem 2.4 *If* $K : \mathfrak{P} \to \mathfrak{T}$ *is rich*, *then* $\bar{F}$ *is the right Kan extension of* F *along* K_1 *and the right Kan extension of* $\tilde{F}$ *along* T.

Our final example of the universal aspect of shape theory is concerned with Grothendieck's notion of a pro-category. Let $\mathfrak{C}$ be a category and let $F : I \to \mathfrak{C}$, $G : J \to \mathfrak{C}$ be functors on (small) cofiltering categories I,J to $\mathfrak{C}$, then F,G are objects of the category Pro-$\mathfrak{C}$, and

$$\text{(2.1)} \qquad \text{Pro-}\mathfrak{C}(F,G) = \varprojlim_{j \in J} \varinjlim_{i \in I} \mathfrak{C}(Fi,Gj)$$

Now let $X\downarrow K$ be the comma category of $\mathfrak{P}$-objects under X, $X \in |\mathfrak{T}|$; and let $D_X : X\downarrow K \to \mathfrak{P}$ be the underlying functor given by

$$D_X(P,f) = P, \text{ where } f : X \to KP$$

$$D_X u = u, \text{ where } u : (P,f) \to (Q,g) \text{ in } X\downarrow K, \text{ that is,}$$

$$u : P \to Q \quad \text{and} \quad Ku \circ f = g$$

Then, as observed independently by K. Morita, in the original restricted context,

$$\text{(2.2)} \qquad \text{Pro-}\mathfrak{P}(D_X,D_Y) \cong \mathfrak{S}(X,Y)$$

However, one may show [3] that (2.2) continues to hold, virtually in complete generality. First we may take (2.1) as the definition of the pro-category even where the index categories (domains of F,G) are no longer cofiltering. This frees us of the necessity, in (2.2), of assuming - or, in any particular case such as the original context, proving - that the comma categories are cofiltering; and then (2.2) is universally true. Thus shape may, in general, be subsumed in the theory of (generalized) pro-categories.

3. Shape, localization and completion

Suppose now that $K : \mathfrak{P} \to \mathfrak{T}$ has a left adjoint $L : \mathfrak{T} \to \mathfrak{P}$. If $\eta : 1 \to KL$ is the unit of the adjunction, we may define a function $\Gamma : \mathfrak{S}(X,Y) \to \mathfrak{T}(X,KLY)$ by the rule

(3.1) $$\Gamma(\tau) = \tau^{LY}(\eta_Y)$$

Let Γ' consist of the composition of Γ and the adjunction-bijection $\mathfrak{T}(X,KLY) \cong \mathfrak{P}(LX,LY)$. One may then prove

<u>Theorem 3.1</u> $\Gamma' : \mathfrak{S}(X,Y) \to \mathfrak{P}(LX,LY)$ <u>is bijective and respects identities and composition</u>. <u>Thus</u> $\mathfrak{S}$ <u>is isomorphic to the Kleisli category of</u> $\mathfrak{T}$ <u>with respect to the triple generated by the adjunction</u> $L \dashv K$.

This theorem implies that, when K admits a left adjoint L, then we may regard a shape morphism from X to Y as an ordinary ($\mathfrak{T}$-)morphism from X to KLY. Moreover given a shape morphism from X to Y, i.e., $f : X \to KLY$, and a shape morphism from Y to Z, i.e., $g : Y \to KLZ$, we compose them, to produce a morphism $h : X \to KLZ$ by the rule

$$h = Kg' \circ f,$$

where g' corresponds to g under the adjunction-bijection $\mathfrak{T}(Y,KLZ) \cong \mathfrak{P}(LY,LZ)$.

As an example of this theorem, consider the following. Let P be a family of prime numbers, let $\mathfrak{T}$ be the category $\mathfrak{N}$ of nilpotent groups and let $\mathfrak{P}$ be the full subcategory $\mathfrak{N}_P$ consisting of P-local nilpotent groups. Then it is known (see, e.g. [8,9]) that the full embedding $K : \mathfrak{N}_P \to \mathfrak{N}$ has a left adjoint L. It is customary to write G_P for LG (or KLG), $G \in |\mathfrak{N}|$, so that

$$\mathfrak{S}(G,H) = \mathrm{Hom}(G,H_P).$$

The localizing map $e : H \to H_P$ is the unit of the adjunction so that a homomorphism $\psi : H \to K_P$ in $\mathfrak{N}$ determines a unique

$\psi_P : H_P \to K_P$ such that $\psi_P e = \psi$. Then we compose $\varphi \in \mathfrak{S}(G,H)$, $\psi \in \mathfrak{S}(H,K)$, that is, $\varphi : G \to H_P$, $\psi : H \to K_P$, to produce $\psi_P \varphi : G \to K_P$.

The notion of richness again enters the story at this point. For one may prove

<u>Proposition 3.2</u> <u>Let</u> $K : \mathfrak{P} \to \mathfrak{T}$ <u>admit the left adjoint</u> $L : \mathfrak{T} \to \mathfrak{P}$. <u>Then the following statements are equivalent</u>:

(i) <u>The triple generated by the adjunction is idempotent</u>;

(ii) <u>K is rich</u>;

(iii) <u>L is rich</u>.

It follows that, if K is rich, then, for all Y in $|\mathfrak{T}|$, KLY is the Adams completion of Y with respect to the morphisms of $\mathfrak{T}$ inverted by L, thus

$$\mathfrak{T}[\Sigma_L^{-1}](-,Y) \cong \mathfrak{T}(-,KLY)$$

Combining this with Theorem 3.1 we have

<u>Theorem 3.3</u> <u>If</u> $K : \mathfrak{P} \to \mathfrak{T}$ <u>is a rich functor admitting a left adjoint</u> L, <u>then</u> $\mathfrak{S} \cong \mathfrak{T}[\Sigma_L^{-1}]$, <u>where</u> Σ_L <u>is the family of morphisms inverted by</u> L.

Now it is easy to see that the family of morphisms of $\mathfrak{T}$ inverted by L coincides with the family of morphisms of $\mathfrak{T}$ inverted by $T : \mathfrak{T} \to \mathfrak{S}$. It is thus reasonable to propose the following question.

<u>Question</u> Suppose $K : \mathfrak{P} \to \mathfrak{T}$ is rich. When is $\mathfrak{S}$ the category of fractions $\mathfrak{T}[\Sigma_T^{-1}]$, where Σ_T is the family of morphisms inverted by T?

It is interesting to note that, when the answer is affirmative, and when the Adams completion Y_T of Y in $|\mathfrak{T}|$ exists, then $\mathfrak{S}(X,Y) = \mathfrak{T}(X,Y_T)$. Thus we are motivated to look for examples (when K does <u>not</u> admit a left adjoint) when the shape morphisms from X to Y are ordinary morphisms from X to some appropriate "modification" of Y.

Bibliography

1. K. Borsuk, Concerning homotopy properties of compacta, Fund. Math. 62 (1968), 223-254.
2. A. Deleanu and P. Hilton, On the categorical shape of a functor, Fund. Math. (1976) (to appear)
3. A. Deleanu and P. Hilton, Borsuk shape and a generalization of Grothendieck's definition of pro-category, Math. Proc. Cam. Phil. Soc. 79 (1976), 473-492
4. A. Dold, Lectures on Algebraic Topology, Springer Verlag (1972)
5. D. Edwards and R. Geoghegan, Shapes of complexes, ends of manifolds, homotopy limits and the Wall obstruction, (1974) (preprint).
6. R. Geoghegan and R. Lacher, Compacta with the shape of finite complexes, Fund. Math. (1975) (to appear).
7. R. Geoghegan and R. Summerhill, Concerning the shapes of finite-dimensional compacta, Trans. Amer. Math. Soc. 179, (1973), 281-292.
8. P. Hilton, Localization and cohomology of nilpotent groups, Math. Zeits. 132 (1973), 263-286.
9. P. Hilton, G. Mislin and J. Roitberg, Localisation of nilpotent groups and spaces, Mathematics Studies 15, North-Holland (1975)

10. G. Kozlowski, Polyhedral limitations on shape (1974) (preprint).

11. J. LeVan, Shape Theory, Dissertation, University of Kentucky (1973).

12. S. MacLane, Categories for the Working Mathematician, Springer-Verlag (1971).

13. S. Mardešić, Shapes for topological spaces, General Topology and its Applications, 3 (1973), 265-282.

14. S. Mardešić and J. Segal, Shapes of compacta and ANR-systems, Fund. Math. 72 (1971), 41-59.

15. S. Mardešić and J. Segal, Equivalence of the Borsuk and the ANR-system approach to shape, ibid., 61-68.

16. W. Holsztynski, An extension and axiomatic characterization of Borsuk's theory of shape, Fund. Math. 70 (1971), 157-168.

17. T. Porter, Generalized shape theory, Proc. Roy. Irish Acad. 74 (1974), 33-48.

Department of Mathematics, Syracuse University
Department of Mathematics, Case Western Reserve University, Battelle Research Center, Seattle, and Forschungsinstitut für Mathematik, ETH, Zürich.

NEW RESULTS IN UNIFORM TOPOLOGY

V.A.Efremovič (Jaroslavl) and A.G.Vaĭnšteĭn (Moscow)

0. This paper contains a short review of some achievements attained in the last 5 years in uniform topology (or proximity geometry) of geodesic spaces by a group of mathematicians of Moscow, Voronež, Gorky and Jaroslavl. We remind that a geodesic space (g.s) (X, ρ) is a complete metric space such that every two points $x, y \in X$ may be connected by a rectilinear segment (i.e., an isometric image of a closed interval in R).

These results may be naturaly grouped into three directions (which are, however, closely connected):

1. Proximity invariants of geodesic spaces.
2. Extension of equimorphisms (i.e., uniform isomorphisms) of geodesic spaces.
3. Applications of uniform topology to differentiable dynamics.

Our main point is to describe the results and to discuss the related unsolved problems, while the proofs will be merely outlined.

Most of the results listed below being obtained in the closest co-operation with D.A.DeSpiller, L.M.Lerman, É.A.Loginov and E.S.Tihomirova, we are deeply grateful to them.

1. Proximity Invariants in Geodesic Spaces

That is one of the oldest problems in proximity geometry of g.s. We cannot even list the main results obtained here and only refer to papers [1 - 9]. But we would like to mention two circumstances concerning this topic; first, that most of the proximity invariants known are, in fact, invariants of the uniform homotopy type (u.h.t., see[7,9 Second, that these invariants have been successfully used not only to distinguish homeomorphic and not equimorphic g.s., but also to solve some problems concerning uniform retraction [8] and behaviour of equimorphism at infinity [5,6] .

2. Extension of Equimorphism of Geodesic Spaces

Motivation. The investigation of this topic has been started on the base of the results concerning the behaviour of equimorphisms of the n-dimensional Lobačevsky space Λ^n (see [10]):

Theorem 2.1. Let B^n be an n-dimensonal ball representing the Poincaré model of Λ^n. Then every equimorphism of Λ^n can be extended to a homeomorphism of the closed n-ball $\bar{B}^n$.

Here we shall discuss a recent generalisation of this theorem [11-13] .

Remarks. a) The points of the Poincaré sphere at infinity S^{n-1} ($= \bar{B}^n \setminus B^n$) are naturally identified as rectilinear rays starting at the origin 0 of the Poincaré model B^n. Below we shall also deal with the constructions using rectilinear rays.
b) Let us consider two uniformities in B^n : one of them may be called the standard uniformity and corresponds to the inclusion $B^n \subset \bar{B}^n$, while the other is defined by the Poincaré metric

$$ds^2 = (1 + r^2)^{-2} (dr^2 + r^2 d\theta^2)$$

and will be called the Lobačevsky uniformity. Thus Theorem 2.1 is equivalent to the following

Theorem 2.1′. Every homeomorphism of B^n which is an equimorphism relative to Lobačevsky uniformity, has the same property relative to the standard uniformity.

This approach suggests the follwoing unsolved

Problem: To describe effectively all pre-compact uniformities compatible with the topology of B^n which may replace the standard uniformity in Theorem 2.1 .

To generalize Theorem 2.1 we shall need the following notions.

Fiberings and Compactifications. Let X be a g.s. such that every bounded subset of X is totally bounded. Let us suppose that, for a certain open bounded $Q \subset X$, $X \setminus Q$ can be fibered into a set Ξ of rectilinear rays ξ in the following sense: Ξ is given a compact topology and the map $x \mapsto (\xi, h)$ is a homeomorphism between $X \setminus Q$ and $\Xi \times R^+$, where $x \in X \setminus Q$, $\xi = px$ is the ray of the fibering running through x and $h = h(x)$ is the distance between x and the starting point of ξ .

<u>Definition 2.2.</u> A sequence $x_n \in X$ will be called <u>Ξ-directed to the ray</u> $\xi \in \Xi$ iff $h(x_n) \to \infty$ and $px_n \to \xi$.

<u>Definition 2.3.</u> A <u>compactification</u> $\bar{X}_{\Xi}$ <u>corresponding to a fibering</u> Ξ is the space $X \cup \Xi$, the topology of this union being defined by convergence of Ξ-directed sequences.

<u>Definition 2.4.</u> A fibering Ξ in a g.s. X will be called <u>exact</u> iff for every other fibering Ξ' in X the identity map $id_X : X \to X$ may be extended to a continuous map $\tau : \bar{X}_{\Xi'} \to \bar{X}_{\Xi}$.

Obviously $\bar{X}_{\Xi_1}$ and $\bar{X}_{\Xi_2}$ are homeomorphic if Ξ_1 and Ξ_2 are both exact; therefore any compactification of X corresponding to an exact fibering will be denoted by $\bar{X}$.

<u>Remark:</u> It is easy to see that Definition 2.4 is equivalent to the following condition: for every fibering Ξ' in X every Ξ'-directed sequence is also Ξ-directed.

<u>Examples of Fiberings.</u> 1) Let $X = M^k$ be a complete simply connected Riemannian manifold of non-positive 2-curvature, $Q \subset X$ an arbitrary ball. We can obtain <u>an exact fibering</u> Ξ in X consisting of all rectilinear rays orthogonal to ∂Q . $\bar{X}_{\Xi}$ is obviously homeomorphic to the k-ball B^k .

2) Let $X = \Lambda^2$ be the Lobačevsky plane, $Q \subset \Lambda^2$ be a bounded open subset with C^1-smooth boundary which consists of two horocycle arcs, belonging to a pair of different horocycles orthogonal to the same straight line in Λ^2 , and connected by a pair of circle arcs. The fibering Ξ consisting of all the rays ξ orthogonal to ∂Q is not exact.

<u>Equivalent and Separated Sequences. Sufficient Conditions for Exactness of a Fibering.</u> Since Ξ is compact, it posesses the only uniformity $\mathcal{U}$ compatible with its topology.

<u>Definition 2.5.</u> Let Ξ be a fibering in a g.s. X, $x_n, y_n \in X$ two sequences going to infinity when $n \to \infty$. We will call them <u>Ξ-equivalent</u> ($x_n \sim y_n$ rel. Ξ) iff $\forall U \in \mathcal{U}\ \exists N \in Z^+ : \forall n > N\ (x_n, y_n) \in U$.

If, for every increasing sequence of integers n_k, $x_{n_k} \not\sim y_{n_k}$ rel.

Ξ , then we will call them Ξ-<u>separated</u> ($x_n \wr y_n$ rel. Ξ).

The following statement is quite obvious.

<u>Proposition 2.6.</u> Let Ξ be an exact fibering and Ξ' an arbitrary one. Then a) $x_n \sim y_n$ rel. Ξ' implies $x_n \sim y_n$ rel. Ξ ; b) $x_n \wr y_n$ rel. Ξ implies $x_n \wr y_n$ rel. Ξ' .

Thus we shall call two sequences <u>equivalent</u> if they are <u>equivalent relative to any exact fibering.</u>

Most of the results listed below are based upon the following simple but important

<u>Lemma 2.7.</u> Let Ξ be an arbitrary fibering in a g.s. X, x_n, $y_n \in X$, $x_n \sim y_n$ rel. Ξ . Then <u>the entire rectilinear segment</u> $x_n y_n$ <u>tends to infinity with</u> n , i.e. every compactum $K \subset X$ intersects only a finite number of these segments.

<u>Sketch of the proof.</u> Suppose $x_n \sim y_n$ rel. Ξ , $z_n \in \overline{x_n y_n}$, $h(z_n) \leq \leq c \;\forall\, n$. Let $d = \operatorname{diam} \{\Xi \times \{0\}\}$, $\xi_n = px_n$, $\eta_n = py_n$. Then for some n the distance $\rho((\xi_n, c + d + 1),(\eta_n, c + d + 1)) < 1$. Obviously $\rho(x_n, y_n) \geq h(x_n) + h(y_n) - 2(c + d)$, but, on the other hand, $\rho(x_n, y_n) < h(x_n) + h(y_n) - 2(c + d) - 1$.

Lemma 2.7 implies the following

<u>Theorem 2.8.</u> a) If a fibering Ξ in a g.s. X satisfies the following <u>Condition</u> A , then Ξ is exact.

<u>Condition A .</u> For every pair of sequences x_n, y_n, $x_n \wr y_n$ rel. Ξ , there exists a fixed ball $K \subset X$ such that $\overline{x_n y_n} \cap K \neq \phi$.

b) If some fibering Ξ in X satisfies Condition A , then any exact fibering in X also satisfies the same condition.

Thus we may say that the space X satisfies Condition A , if any fibering in X satisfies this condition.

Geometry implied by Condition A may be illustrated by the following

<u>Theorem 2.9.</u> If a g.s. X satisfies Condition A , then every two points $x, y \in \overline{X}$ can be connected by a rectilinear path (relative to the metric ρ).

Sketch of the proof. The only case to study is that of two points $x, y \in \overline{X} \setminus X$. Let $x_n, y_n \in X$, $x_n \to x$, $y_n \to y$ in $\overline{X}$. Then $x_n \sim y_n$ rel. Ξ and thus Condition A implies the existence of a point $z_n \in x_n y_n$ such that $h(x_n)$ is bounded. It is easy to choose an increasing sequence of integers n_k such that $\overline{x_{n_k} y_{n_k}}$ tends to a straight line $\Gamma \subset X$ linking x with y in $\overline{X}$.

Examples. 3) The Lobačevsky space Λ^n satisfies Condition A ; Theorem 2.9 describes the limiting line effect.

4) Let X be a surface of revolution with the Riemannian metric given by the formula

$$ds^2 = dr^2 + f^2(r)\, d\Theta^2,$$

$$0 \leq \Theta \leq 2\pi \ , \quad r \in R^+ \ ;$$

a) if $f(r) = r^a$, $a \geq 1$, than X satisfies Condition A iff $a > 1$;

b) if $f(r) = r \log r$, then X satisfies Condition A .

The Notion of a λ-Chain. Equimorphisms of Spaces with Exact Fiberings. To describe the behaviour of the image of a rectilinear ray under an arbitrary equimorphism, we shall introduce the following

Definition 2.10. Let $\lambda \in R$, $\lambda > 0$, $x, y \in X$. A finite set $Z = \{z_j \in X \mid 0 \leq j \leq m\}$ will be called a λ-chain, linking x with y, iff

a) $z_0 = x$, $z_m = y$; b) $1 \leq \rho(z_{j-1}, z_j) < 2$; c) $\sum_{j=k}^{1} \rho(z_{j-1}, z_j) \leq$

$$\leq \lambda \rho(z_{k-1}, z_1) \qquad \forall\, k, 1 : 1 \leq k \leq 1 \leq m \ .$$

The properties of uniformly continuous maps proved in [1] imply the following important

Lemma 2.11. Let X be a g.s., $f : X \to X$ an equimorphism. Then there exist $\lambda \geq 1$, $c > 0$ such that, for every $x, y \in X$, either $\rho(x,y) < c$ or there exists a λ-chain $Z \subset f(\overline{xy})$ linking $f(x)$ with $f(y$

Now we shall strengthen Condition A replacing rectilinear segments by λ-chains. Thus we obtain the following

Condition B. For an arbitrary $\lambda \leq 1$ and for an arbitrary sequence of λ-chains $Z_n = \{x_n = z_{n0}, z_{n1}, \dots, z_{nm_n} = y_n\}$ such that $x_n \sim y_n$, there exists a fixed ball $K \subset X$ which intersects all these chains.

Remark. Obviously Condition A may be replaced by Condition B in Theorem 2.8.

Now we shall formulate the central results of this section.

Theorem 2.12. If a g.s. X satisfies Condition B, then every equimorphism $f : X \to X$ preserves the equivalence relation between the sequences in X.

Theorem 2.13. If a g.s. X satisfies Condition B, then every equimorphism $f : X \to X$ can be extended to a homeomorphism $\bar{f} : \bar{X} \to \bar{X}$.

Both these theorems are obviously implied by Lemmas 2.7 and 2.11. We shall call stable a g.s. X satisfying the conclusion of Theorem 2.13.

A Metric Condition for Stability. To check whether Condition B is satisfied, we shall introduce the following construction. Let $d(\cdot, \cdot)$ be any metric on Ξ compatible with its topology. We determine a function $\mu : R^+ \to R^+$ by the following formula:

$$\mu(t) = \inf \{ \rho(x,y) / d(\xi, \eta) \mid \xi = px, \ \eta = py, \ h(x) \geq t, \ \rho(x,y) \geq 2 \} . \tag{3}$$

Theorem 2.14. If the infinite integral

$$J = \int_1^\infty (\mu(t)^{-1}) dt \tag{4}$$

converges, then Ξ satisfies Condition B.

Sketch of the proof. Suppose that $Z_n = \{x_n = z_{n0}, \dots, z_{nm_n} = y_n\}$ is a sequence of λ-chains, Z_n goes to infinity with n, $x_n \sim y_n$ rel. Ξ. Let us denote $\xi_n = px_n$, $\eta_n = py_n$, $h_n = h(x_n)$, $h_n^k = (z_{nk})$, $0 \leq k \leq m_n$, $L = 3\lambda + 1$; we may suppose that $h_n \leq h_n^k$. For all n sufficiently large

$$h_n^k \geq (h_n + k)/ L \quad ; \qquad (5);$$

on the other hand

$$d(\xi_n, \eta_n) \leq \sum_{k=1}^{m_n} \rho(z_{n,k-1}, z_{nk}).(\mu(h_n^k - 2))^{-1} ; \qquad (6);)$$

since μ does not decrease (5) and (6) imply

$$d(\xi_n, \eta_n) \leq \int_{h_n/L-2}^{\infty} (\mu(t))^{-1}\, dt ;$$

thus $d(\xi_n, \eta_n) \to 0$, since J converges.

Examples. 5) Λ^n satisfies Condition B since μ grows exponentially; this proves Theorem 2.1.

6) Let X be a surface of revolution described in Example 4 a); if $a > 1$ then X satisfies Condition B .

7) Let X be a surface revolution described in Example 4 b) ; it is easy to see that the map $f : X \to X$ given by the formula

$$f(r, \theta) = \begin{cases} (r, \theta), & r \leq e \\ (r, \theta + \log\log r), & r > e \end{cases}$$

is an equimorphism of X . Since f cannot be extended to $\bar{X}$, X does not satisfy Condition B ; thus A does not imply B .

Exactness of the "Central" Fibering. To prove that the "central" fibering of Example 1 is exact, we shall introduce a new condition of exactness.

Theorem 2.15. If a fibering Ξ in a g.s. X satisfies the following Condition C , then Ξ is exact.

Condition C. For a fibering Ξ in a g.s. X

(C1) If $x_n, y_n \in X$, $h(x_n) \to \infty$, $\rho(x_n, y_n) <$ const, then

$x_n \sim y_n$ rel. Ξ.

(C2) Every rectilinear ray in X is Ξ-directed.

(C3) If a rectilinear ray $\Gamma \subset X$ is Ξ-directed to a $\xi \in \Xi$ then the distance between a point $x \in \Gamma$, going to infinity along Γ, and ξ does not increase.

The classical results by J.Hadamard and E.Cartan obviously imply that the "central" fibering satisfies Condition C.

The proof of Theorem 2.15 is based upon the following trivial

Lemma 2.16. Let Ξ be an arbitrary fibering in a g.s. X, $x_n \in X$, let x_n be Ξ-directed to $\xi \in \Xi$, $\xi_n = px_n$. Then $\forall z \in \xi, \forall \varepsilon > 0$ $\exists N = N(z, \varepsilon) : \forall n > N \quad \exists y_n \in \xi_n : \rho(y_n, z) < \varepsilon$.

Sketch of the proof of Theorem 2.15. Suppose that Ξ and Ξ' are two fiberings in a g.s. X, Ξ satisfies Condition C, $x_n \in X$, x_n is Ξ-directed to $\xi' \in \Xi'$, $\xi'_n = p'x_n$; then ξ'_n, ξ' are Ξ-directed to $\xi_n, \xi \in \Xi$, respectively. Using Lemma 2.16 we may find a sequence $y_n \in \xi'_n$, $h(y_n) \to \infty$, such that $\rho(y_n, \xi') \to 0$. (C3) implies that $\rho(y_n, \xi) <$ const for $\forall n$. Let $u_n, v_n \in \xi_n$ be the nearest points to x_n, y_n, respectively, and let $w_n \in \xi$ be the nearest point to y_n. Then $y_n \sim w_n$ rel. Ξ, $x_n \sim u_n$ rel. Ξ and $v_n \sim y_n$ rel. Ξ. Since obviously $u_n \sim v_n$ rel. Ξ we have $x_n \sim w_n$ rel. Ξ.

Remark. We do not know if Theorem 2.8 b) may be extended to Condition C; this is true in any complete, simply connected Riemannian manifold of non-positive 2-curvature.

Adjoining Results. These concern mainly two problems: a) equimorphisms of Λ^n and strong rigidity theorems, see [9, 14-16]; b) equimorphisms of R^n, see [17], where also some "rigidity theorems" may be obtained. We shall just mention one of them which is in fact rather trivial.

Set M_α, $\alpha \in R$, a 3-manifold with locally Euclidean Riemannian metric obtained from $\{(x,y,z) \in R^3 \mid 0 \le z \le 1\}$ by identifying $(x,y,0)$ with $(x \cos\alpha - y \sin\alpha,\ x \sin\alpha + y \cos\alpha,\ 1)$

Theorem 2.17. M_α and M_β are equimorphic iff $|\alpha| \equiv |\beta| \mod 2\pi$.

3. Applications of Uniform Topology to Differentiable Dynamics

Results listed below have been obtained in collaboration with L.M.Lerman.

Motivation. The problem has been suggested by the new approach developed by L.M.Lerman and L.P.Šilnikov [18]. To study time-dependent vector fields on a compact smooth manifold M, they considered $M \times R$ as posessing the Cartesian product uniformity. Using this uniformity and the 1-foliation into integral curves of a time dependent vector field $\vec{v}(x,t)$, $x \in M$, $t \in R$, $\vec{v} \in T_x M$, most of the qualitative properties of such a vector field may be expressed. The triple ($M \times R$, Cartesian product uniformity, 1-foliation into integral curves of a time-dependent bounded vector field $\vec{v}$) will be called an integral portrait of $\vec{v}$. Equivalence of two vector fields has been defined in such a way, and some structural stability theorems have been proved.

Here uniformity is essential, since any time dependent vector field generates an integral portrait topologically equivalent to the trivial 1-foliation $L : M \times R = \bigcup_{x \in M} (\{x\} \times R)$.

Let M be a compact C^∞-manifold, $f \in \mathrm{Diff}^1(M)$. We are going to produce a triple (topological space $M \times R$, a certain compatible uniformity in $M \times R$, 1-foliation L) such that most of the qualitative properties of f may be studied using this triple. All we have to do is to define a compatible uniformity in $M \times R$. Define $\hat{f} : M \times R \to M \times R$ by the following formula:

$$\hat{f}(x,t) = (f(x),\ t-1),\quad x \in M,\quad t \in R .$$

Lemma 3.1. There exists a weakest compatible uniformity $\mathcal{U}_f$ on $M \times R$ such that $\hat{f}$ generates a uniformly equicontinuous group relative to $\mathcal{U}_f$.

<u>Sketch of the proof.</u> $M \times R / \{ z = f(z) \}$ is a compact C^∞-manifold M_f, and the natural projection $M \times R \to M_f$ is a (smooth) covering. The only compatible uniformity in M_f can be naturally <u>lifted up to</u> $M \times R$ to produce the desired uniformity $\mathcal{U}_f$.

<u>Remark.</u> We can easily see that $\mathcal{U}_f$ is determined by a certain C^∞ Riemannian metric on $M \times R$.

Later the pair $(M \times R, \mathcal{U}_f)$ will be abbreviated to $\widetilde{M}_f$ and the triple $(M \times R, \mathcal{U}_f, L)$ to the pair $(\widetilde{M}_f, L_f)$. $\widetilde{M}_f$ will be called a non-autonomous suspension over f.

<u>Definition 3.2.</u> $f, g \in \mathrm{Diff}^1(M)$ are called δ-<u>equivalent</u> iff there exists an equimorphism $\Phi: \widetilde{M}_f \to \widetilde{M}_g$ such that $\Phi(L_f) = L_g$.

This definition is motivated by the following trivial

<u>Proposition 3.3.</u> Two leaves $L(x), L(y) \in L_f$ such that $L(x) \ni (x,0)$, $L(y) \ni (y,0)$, are in proximity relative to $\mathcal{U}_f$ iff the same holds for the orbits of x and y under f relative to the only compatible uniformity on M.

<u>Classification of Non-Autonomous Suspensions.</u> Now it seems quite natural to suggest the following

<u>Problem:</u> When are M_f and M_g, f,g $\mathrm{Diff}^1(M)$, equimorphic? This problem may be considered as directly related to those of the first section of this paper, but we have described above its relation to dynamics.

To illustrate the problem, let us consider some

<u>Examples.</u> 1) Let $M = S^n$ be an n-dimensional sphere. Since the group Homeo (S^n) consists of only two path connection components, every $f \in \mathrm{Diff}^1(M)$ gives a non-autonomous suspension equimorphic to $S^n \times R$.

2) Let M be an arbitrary smooth compact manifold, and let f generate an equicontinuous group of homeomorphisms of M. Then $\widetilde{M}_f$

is equimorphic to $M \times R$.

Remarks: a) It is easy to prove that $f \in \mathrm{Diff}^1(M)$ is δ-equivalent to id_M iff f generates an equicontinuous group.
b) The previous remark is of special interest when compared to Theorem 2.17.
3) Let $M = T^2$ be a 2-torus, f_0, f_1, f_2, f_3 its linear diffeomorphisms corresponding to matrices:

$$\begin{pmatrix} 1 & 0 \\ 0 & 1 \end{pmatrix}, \begin{pmatrix} 1 & 1 \\ 0 & 1 \end{pmatrix}, \begin{pmatrix} 2 & 1 \\ 1 & 1 \end{pmatrix}, \begin{pmatrix} 3 & 2 \\ 1 & 1 \end{pmatrix} .$$

$\widetilde{M}_{f_j}$ will be abbreviated here to $\widetilde{M}_j$. To distinguish $\widetilde{M}_j$ neither uniform homology [3,4] nor the volume invariant [1] are sufficient. Using the ideas of [2] one may check, however, that $\widetilde{M}_i$ is not equimorphic to $\widetilde{M}_j$, $i = 0,1$, $j = 2,3$. But this is not sufficient to distinguish $\widetilde{M}_0$ from $\widetilde{M}_1$ and $\widetilde{M}_2$ from $\widetilde{M}_3$.

Uniform Homotopy Type of Non-Autonomous Suspensions. Luckily, uniform homotopy type of non-autonomous suspensions over diffeomorphisms can be studied in rather an explicit way (see [19,20]).

Theorem 3.4. $\widetilde{M}_f, \widetilde{M}_g$ have the same u.h.t. iff there exist $n, m \in Z \setminus \{0$ $\varphi : M \to M$ a homotopy equivalence, such that $\varphi \circ f^m$ and $g^n \circ \varphi$ are homotopic.

The proof will be outlined below; first we shall list some corollaries.

Corollary 3.5. If $\widetilde{M}_f$ is equimorphic to $M \times R$, then there exists a $k \in Z \setminus \{0\}$ such that f^k is homotopic to the identity map.

Corollary 3.6. If $f \in \mathrm{Diff}^1(M)$ is Anosov (see [21]), then $\widetilde{M}_f$ is not equimorphic to $M \times R$.

This corollary is implied by the previous one and the following important

Theorem 3.7. No Anosov diffeomorphism of a compact manifold is homotopic to the identity map.

The proof of this theorem belongs completely to algebraic topology and is therefore omitted here.

Corollary 3.8. Let $M = T^s$ be an s-torus, $f,g \in GL(s,Z)$ its linear diffeomorphisms. Then T^s_f and T^s_g are equimorphic iff there exist $n,m \in Z\setminus\{0\}$ and $h \in GL(s,Z)$ such that $h \circ f^m = g^m \circ h$.

To prove this corollary one should apply Theorem 3.4 and consider the natural action of f,g,φ in the fundamental group of M. Notice, that this corollary implies the non-existence of an equimorphism between $\widetilde{M}_i$ and $\widetilde{M}_j$, $i \neq j$, described above in Example 1.

The purely algebraic problem, involved now with the uniform classification of non-autonomous suspensions over linear diffeomorphisms of tori, seems rather difficult, and we do not know the complete solution even for $s = 2$!

Corollary 3.8 may be easily generalized to the algebraic diffeomorphisms of infra-nilmanifolds (see [22]). Thus the problem of uniform classification of non-autonomous suspensions over all Anosov diffeomorphisms, known to us, is reduced to a purely algebraic problem (cf.[23]).

Sketch of the Proof of Theorem 3.4. The proof of Theorem 3.4 is based upon two propositions listed below.

Lemma 3.9. Let K be a compactum, N a Riemannian manifold, A an arbitrary infinite set of indices, $\{f_\alpha : K \to N \mid \alpha \in A\}$ a precompact set in $C(K,N)$ (relative to the usual topology). Then there exists $\alpha^* \in A$ such that for an infinite subset $B \subset A$ f_{α^*} and f_β are homotopic whenever $\beta \in B$.

This is quite obvious, but implies some interesting corollaries, e.g.

Corollary 3.10. Let M be a compact Riemannian manifold, and let $f \in H_omeo(M)$ generate an equicontinuous group. Then f^k is homotopic to id_M for some integer $k \neq 0$.

Lemma 3.11. Let $f,g \in Diff^1(M)$, $\Phi : M_f \to M_g$ be a uniform homotopy (u.h.) equivalence, $M_0 = M \times \{0\} \subset M_f$. Then there exists a sequence of integers $l(k)$ such that

a) $\hat{g}^{l(k)} \circ \Phi \circ f^k|_{M_o} : M \to \tilde{M}_g$ is a precompact set of mappings, $k \in Z$; b) there exists $L \geq 1$ such that $L^{-1} \leq |l(k)/k| \leq L$.

To prove this lemma one should use the properties of u.h. equivalences of g.s. described in [9] ; these properties imply that $l(k)$ may be defined by the formula $(M \times \{t\} \subset \tilde{M}_g)$:

$$l(k) = - \text{entier} (\inf \{t \mid M \times \{t\} \cap \Phi \circ \hat{f}^k(M_o) \neq \phi\}).$$

As far as the necessity of its conditions is concerned, Theorem 3.4 is obviously implied by the two lemmas above. Sufficiency is obvious.

Adjoining Results. These concern mainly necessary conditions for two diffeomorphisms to be δ-equivalent. We shall mention two theorems to illustrate this approach.

Theorem 3.12. Let $f,g \in \text{Diff}^1(M)$, and let f and g be δ-equivalent; if topological enthropy of f is 0 , then the same holds for g .

Theorem 3.13. Under the assumptions of the above theorem there exist integers $m,n \neq 0$ and $\varphi \in \text{Homeo}(M)$ such that $\varphi^{-1} \circ g^{-n} \circ \varphi \circ f^m$ belongs to the path connection component of id_M in Homeo (M).

We hope to achieve further results in this direction involving structural stability of wider classes of time dependent diffeomorphisms and vector fields.

References

[1] V.A.Efremovič: Proximity Invariants,(Russian) Ivanov.Gos. Ped. Inst.Učen. Zap. 31 (1963), vyp. mat.,74-81. MR 47# 9505.

[2] A.S.Švarc: A volume invariant of coverings. Dokl. Akad. Nauk SSSR (N.S.) 105 (1955), 32-34. (Russian) MR 1 p. 18 .

[3] E.S.Tihomirova: Uniform homology groups. (Russian) Izv. Akad. Nauk SSSR Ser. Mat. 26 (1962), 865-876. MR 27# 6263.

[4] E.S.Tihomirova: The spectrum of uniform homologies. Dokl. Akad. Nauk SSSR 191 (1970), 1235-1237 (Russian); translated as Soviet Math. Dokl. II (1970), 538-541. MR 44 # 1021.

[5] V.A. Efremovič, E.S.Tihomirova: Volumes and capacities in proximity geometry. (Russian) Dokl. Akad. Nauk SSSR 214 (1974), 29-32. MR 50 # 3167.

[6] E.S.Tihomirova: Invariants of equimorphisms in Riemannian manifolds. All -Union Sci. Conf. in Non-Euclidean Geometry" 150 Years of Lobačevsky Geometry", Abstracts, 195, Kazan (1976).

[7] A.G.Vaĭnšteĭn: Uniform homotopy and proximity invariants.(Russian. English summary) Vestnik Moskov. Univ. Ser. I Mat. Meh. 25 (1970), no. 1, 17-20.MR 43 # 5531.

[8] A.G.Vaĭnšteĭn: Uniform homology group and uniform retracts.(Russian.English summary) Vestnik Moskov. Univ. Ser. I. Mat. Meh.26 (1971), no. 4, 59-62. MR 45 #4407.

[9] A.G.Vaĭnšteĭn: Equimorphisms of geodesic spaces and uniform homotopy. (Russian) Vestnik Moskov. Univ. Ser. I Mat. Meh. 28 (1973), no. 3, 66-69.

[10] V.A.Efremovič, E.S.Tihomirova: Equimorphisms of hyperbolic spaces. (Russian) Izv. Akad. Nauk SSSR Ser. Mat. 28 (1964), 1139-1144. MR 29 # 6374.

[11] V.A.Efremovič, È.A.Loginov, E.S.Tihomirova: Equimorphisms of Riemannian manifolds. (Russian) Dokl. Akad. Nauk SSSR 197 (1971), 25-28. MR 44 # 2171.

[12] V.A.Efremovič, È.A.Loginov: Extension of equimorphisms in Riemannian geometry. (Russian) Uspehi Mat. Nauk 27 (1971), no.6 , 237-238.

[13] A.G.Vaĭnšteĭn, V.A.Efremovič, È.A.Loginov: Infinity in geodesic spaces. (Russian) Dokl. Akad. Nauk SSSR 220 (1975), no.3, 505-508.

[14] D.A.De-Spiller: Equimorphisms, and quasiconformal mappings of the absolute. (Russian) Dokl. Akad. Nauk SSSR 194 (1970), 1006-1009. MR 42 # 6698.

[15] G.A.Margulis: The isometry of closed manifolds of constant negative curvature with the same fundamental group. (Russian) Dokl. Akad. Nauk SSSR 192 (1970), 736-737. MR 42 # 1012.

[16] G.D.Mostow: Strong rigidity of locally symmetric spaces. Princeton Univ. Press, 1973. MR 52 # 5874.

[17] A.G.Vaĭnšteĭn: Uniform classification of isometries in Euclidean and Lobačevsky spaces. (Russian) Uspehi Mat. Nauk 29 (1974),no.5, 217-218,Erratum, ibidem 30 (1975), no. 4, 301.

[18] L.M.Lerman, L.P.Šil'nikov: The classification of structurally stable nonautonomous second order systems with a finite number of cells. (Russian) Dokl. Akad. Nauk SSSR 209 (1973), 544-547. MR 48 # 3077.

[19] A.G.Vaĭnšteĭn, L.M.Lerman: Proximity geometry and non-autonomous suspensions over diffeomorphisms. (Russian) Uspehi Mat. Nauk 31 (1976), no. 5, 231-232.

[20] A.G.Vaĭnšteĭn, L.M.Lerman: Uniform topology and time dependent flows. All-Union Sci.Conf. in Non-Euclidean Geometry " 150 Years of Lobačevsky Geometry " , Abstracts, 38, Kazan (1976).

[21] D.V.Anosov: Geodesic flows on closed Riemannian manifolds of negative curvature. (Russian) Trudy Mat. Inst. Steklov 90 (1967), 209 pp. MR 36 # 7157.

[22] Z.Nitecky: Differential dynamics. MIT Press, 1971.

[23] A.Manning: There are no new Anosov diffeomorphisms on tori. Amer. J. Math. 96 (1974), 422-429. MR 50 # 11324.

TOPOLOGIZATION OF BOOLEAN ALGEBRAS

J. Flachsmeyer

Sektion Mathematik Ernst-Moritz-Arndt Universität
Jahnstrasse 15a, 22 Greifswald, DDR

Introduction

We start with the standard facts in the category of Boolean rings, Boolean lattices and Boolean spaces. Then we define the hyperstonian cover induced by the second function dual of a compact space and show the connections with the Gleason cover. Section 2 contains the notion of a Boolean ring of idempotents of a ring and some relevant examples. Theorem 4 identifies the Stone representation spaces for the Boolean algebra of all bands in the case of the function lattice C(S) and the measure lattice M(X). (Part I of this theorem seems to be foklore (?); but part II seems to be new, for the special case of Boolean spaces being a consequence of the investigations of D. A. EDWARDS). In Section 3 we show how one can obtain compatible Boolean ring topologies by algebra lattice topologies on the function space of the corresponding representation space. An important procedure which yields algebra lattice topologies arises by working with measures. (Theorem 6 on Riesz topologies.) For Boolean algebras the corresponding topologies are the Nikodym topologies (Corollary of Theorem 6). Such topologies on Boolean algebras were studied by J. L. B. COOPER and D. A. EDWARDS. The concluding remarks concern topological semifields in the sense of ANTONOVSKI-BOLTJANSKII-SARYMSAKOV. Our main point of view - the interplay of Boolean ring topologies and function topologies - has here its natural action.

1. Preliminaries

1.1. The Stone functor

For the basic facts concerning Boolean algebra we refer to SIKORSKI [24], HALMOS [16] and SEMADENI [23]. It is possible to develop Boolean theory from several points of view, one of which is of algebraic character (Boolean rings), another one of order-theoretic type (Boolean lattices), while the third one is of topological nature (Boolean spaces). The mutual interplay of these three directions was strongly emphasized by the fundamental work of M. H. STONE [25]. With every Boolean ring a Boolean lattice is associated and vice versa. This correspondence effects an isomorphy between the category of all Boolean rings BoolRg, i.e. the idempotent rings with their ring homomorphisms, and the category of all Boolean lattices BooLat, i.e. the distributive relatively complemented lattices with their lattice homomorphisms preserving the relative complement. In the case of Boolean rings with unit the restriction to such homomorphisms is necessary which preserve the unit. This yields the subcategory BoolRgI. An equivalent lattice category is that of BoolAlg, where the Boolean algebras, i.e. distributive complemented lattices, are the objects and the lattice homomorphisms preserving the complement are the morphisms. The prototype of the natural correspondence of Boolean lattices and Boolean rings is that of a field of sets (=algebra of sets) with the operations of union, intersection and set-complementation and the same family of sets with the operations of symmetric difference and intersection. Every Boolean algebra (resp. lattice) can be regarded as a field of sets (resp. ring of sets). By the important Stone representation theorem to every Boolean algebra (Boolean lattice) $\mathbb{B}$ there corresponds a topologically unique Hausdorff zero-dimensional compact (locally compact) space- the so called spectrum Spec($\mathbb{B}$) of $\mathbb{B}$ - such that $\mathbb{B}$ is isomorphic to the field of all clopen subsets of Spec($\mathbb{B}$) (resp. the ring of all compact-open subsets). The field of all clopen subsets of a Boolean space X (=Hausdorf zero-dimensional compact space) is called the dual algebra $\mathbb{D}$ (X) of X . The Stone representation produces a dual equivalence of the category BoolAlg and the category Comp_0 of Boolean spaces and their continuous maps.

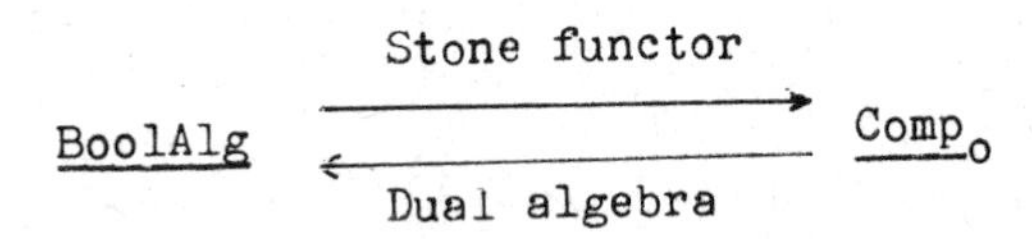

For the case of BoolRg resp. BoolLat it is convenient to exclude the zero-ring resp. the zero-lattice and to make restriction to proper homomorphisms (=non-zero homomorphisms) only. Then the Stone representation yields a dueal equivalence of BoolRg (resp. BoolLat) to the category LocComp_0 of locally compact zero-dimensional spaces and proper maps (=continuous closed maps with compact fibres).

1.2. The hyperstonian cover induced by the second function dual of a compact space

If the Stone functor is used, two Banach lattices can be immediately associated with each Boolean algebra $\mathbb{B}$, the space $C(\mathrm{Spec}(\mathbb{B}))$ of all continuous realvalued functions on $\mathrm{Spec}(\mathbb{B})$ and the space $M(\mathrm{Spec}(\mathbb{B}))$ of all bounded regular Borel measures on $\mathrm{Spec}(\mathbb{B})$. By the Riesz integral representation theorem it is wellknown that for every $X \in$ Comp(=category of all compact Hausdorff spaces and their continuous maps), $M(X)$ is the Banach dual of $C(X)$. $M(\mathrm{Spec}(\mathbb{B}))$ is the same as the Banach lattice $M(\mathbb{B})$ of all bounded finitely additive measures on $\mathbb{B}$ (cf. EDWARDS [9]). While $M(X)$ is always a complete vector lattice (because it is the order dual of $C(X)$), $C(X)$ becomes a complete vector lattice iff X is a Stonian space, i.e. every open subset has open closure (= an extremally disconnected space) (Stone-Nakano theorem [26], [19]). It is possible to assign each $X \in$ Comp two Stonian spaces. The first one is the spectrum of the complete Boolean algebra $R_o(X)$ of all regular open sets of X, moreover, there is a canonical continuous map of $\mathrm{Spec}(R_o(X))$ onto X. This situation is characterized as follows (cf. GLEASON [15]):
For every $X \in$ Comp there is a unique (up to a topological isomorphism) Stonian space gX with a continuous irreducible map onto X (i.e. no proper closed subset is mapped onto X)

$$g:gX \longrightarrow X .$$

This mapping situation (resp. the preimage space gX) will be called the Gleason cover of X. (It is also usual to call it the Gleason resolution or the projective resolution of X). From a more functional analytic point of view the Gleason cover arises as the structure space of the Dedekind-MacNeille completion of the vector lattice $C(X)$ (cf. DILWORTH [6], and for another approach FLACHSMEYER [13]).

The second Stonian space associated with X can be obtained as follows:
The second Banach dual $C''(X)$ of $C(X)$ has the Kakutani representation $C(X)$ as a complete M-space over a Stonian space $\tilde{X}$. The canonical injection of $C(X)$ in its second dual then induces a continuous surjection from $\tilde{X}$ onto X. This mapping situation (resp. the preimage space) will be called the hyperstonian cover of X

$$h:hX \longrightarrow X$$

induced by the second function dual. The space hX is hyperstonian in the sense of DIXMIER [7]. This means that the union of the supports of all hyperdiffuse measures $\nu \in M(hX)$ forms a dense subset of hX. A positive measure $\nu \in M^+(hX)$ is said to be <u>hyperdiffuse</u> (or normal) iff every nowhere dense Borel set has ν-measure zero. In general ν is hyperdiffuse iff ν^+ and ν^- are hyperdiffuse. The canonical embedding of $M(X)$ into its second topological dual $M''(X) \cong M(hX)$ gives precisely the hyperdiffuse measures on hX.

<u>Problem</u>

Is there a nice topological descriptive characterization of the hyperstonian cover $h : hX \to X$ induced by the second function dual of X? This question is far from being answered.
Here we only want to compare the two covers $g : gX \to X$ and $h : hX \to X$. For this reason we give the following new characterization of the projective objects in <u>Comp</u>.

<u>Theorem 1</u>

A space $X \in$ <u>Comp</u> is Stonian iff every continuous surjection $f : Y \to X$ for $Y \in$ <u>Comp</u> is a retraction, i.e. there is an embedding $e : X \to Y$ such that $e \circ f$ is the identity on $e(X)$.

<u>Proof</u>

The extremal disconnectednes follows easily from the retraction property, because under this assumption the Gleason cover $g : gX \to X$ must be a retraction: $e \circ g|_{e(X)} = id_{e(X)}$. By the irreducibility of g the map e must be a topological isomorphism between gX and X. Conversely let X be extremally disconnected and let $f : Y \to X$ be a given epimorphism in <u>Comp</u>. With the help of Zorn's lemma f can be reduced to a closed subspace Y_0 of Y, $f_0 : Y_0 \to X$, $f_0 \equiv f|_{Y_0}$, such that f_0 is irreducible. The projectivity of an extremally disconnected space implies that an irreducible surjection onto such a space has to be a topological isomorphism (see [15] for <u>Comp</u> or [10] for a much greater category). Now let e be the inverse map of f_0 from X onto Y_0. Then $e \circ f$ is the identity on Y_0, thus f is a retraction. □

<u>Theorem 2</u>

Let $f : S \to X$ be a continuous surjection of a Stonian space S onto a given compact Hausdorff space X. Then every restriction $f_0 : S_0 \to X$ of f to an irreducible map f_0 on the closed subspace $S_0 \subset S$ is isomorphic to the Gleason cover.

Proof

Let $f_o : S_o \to X$ be a restriction of the above mentioned type. The projectivity of S yields the mapping diagram

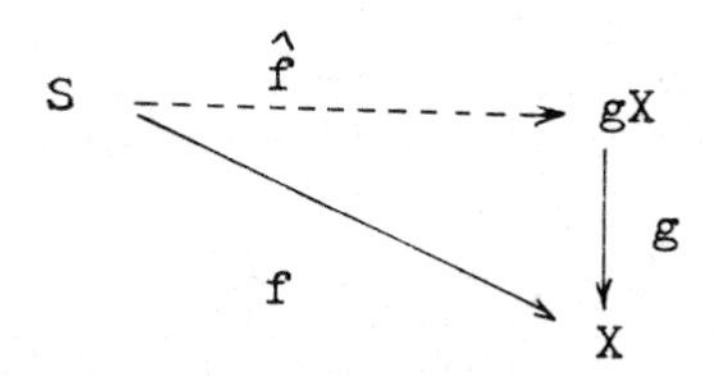

Because of irreducibility of g the map f must be onto (hence, by the preceding theorem, f is a retraction). The irreducible restriction $f_o := f|_{S_o}$ induces an irreducible $\hat{f}_o := \hat{f}|_{S_o}$. Thus $\hat{f}_o$ is a topological isom8rphism.□

Corollary

For every $X \in$ Comp the hyperstonian cover $h : hX \to X$ induced by the second function dual contains the Gleason cover. Any irreducible restriction of h to a closed subspace of hX yields the Gleason cover.

Remark

The situation is clearly arranged for the case $\alpha \mathbb{N}$ - the Alexandrov-one-point compactification of the natural numbers.

The Gleason cover is established by the Stone-Čech-compactification with the natural projection map onto $\alpha \mathbb{N}$. The hyperstonian cover is given by the projection of the Stone-Čech-compactification $\beta \overline{\mathbb{N}}_{dis}$ of the discrete space $\mathbb{N} \cup \{\infty\}$ onto $\alpha \mathbb{N}$. In general, the hyperstonian cover $h : hX \to X$ contains even the part consisting of the natural projection of βX_{dis} onto X. (This part is associated with the discrete measures on X).

2. Some Boolean algebras of functional - analytic character

2.1. Boolean rings of idempotents of a ring

Let $\mathcal{R} = (R,+,\cdot)$ be a (not necessarily commutative) ring. The set $I(\mathcal{R}) = \{x : x^2 = x, x \in R\}$ of all idempotents of $\mathcal{R}$ has the following properties, which are easy to prove:

(1) $0 \in I(\mathcal{R})$.

(2) If $x, y \in I(\mathcal{R})$ and x, y commute, i.e. $xy = yx$, then $x \cdot y \in I(\mathcal{R})$.

(3) If $x, y \in I(\mathcal{R})$ and x, y commute, then $x + y - x \cdot y \in I(\mathcal{R})$, $x + y - 2xy \in I(\mathcal{R})$.

(4) If $\mathcal{R}$ has a unit element 1, then for $x \in I(\mathcal{R})$ also $1 - x \in I(\mathcal{R})$.

(5) A natural partial order can be introduced in $I(\mathcal{R})$ by the following definition: $x, y \in I(\mathcal{R})$, $x \leq y :\Leftrightarrow xy = yx = x$. With respect to this natural partial order, $I(\mathcal{R})$ has the following properties: Every pair x, y of commuting idempotents has an infimum
$\inf(x,y) = xy$
and supremum
$\sup(x,y) = x + y - xy$.

A <u>Boolean ring</u> $\mathbb{B}$ <u>of idempotents of</u> $\mathcal{R}$ is defined to be a nonvoid commuting set $\mathcal{B} \subset I(\mathcal{R})$ which is a Boolean ring with respect to the following operations

$$x \boxplus y = x + y - 2xy ,$$
$$x \boxdot y = xy .$$

The above quoted properties imply by virtue of Zorn's lemma:

(6) Every non-void commuting subset $K \subset I(\mathcal{R})$ is contained in a maximal one and every maximal commuting subset of idempotents forms a Boolean ring of idempotents of $\mathcal{R}$.
If $\mathcal{R}$ is a commutative ring, the set $I(\mathcal{R})$ of all idempotents is a Boolean ring of idempotents.
Special examples of this general procedure to get Boolean rings resp. algebras from (may be non-commutative) rings will be considered in the sequel in a more topological situation.

2.2. <u>Boolean rings of continuous indicator functions</u>

Let X be a topological space. Then the rings $C(X)$ and $C_b(X)$ of all continuous real-valued (resp. bounded c.r.v.) functions on X yield the same Boolean ring of all idempotents. Of course, a function $f \in C(X)$ is an idempotent iff f takes only values 0 or 1 and the support supp f is a clopen set in X .

The representation theorem of Stone can be formulated in the terminology of functions as follows:
For every abstract Boolean ring $\mathbb{B}$ there exists a unique (up to a homeomorphism) locally compact Boolean space X such that $\mathbb{B}$ is isomorphic to the Boolean ring of all idempotents of the ring $C_{oo}(X)$ of all continuous real-valued functions with compact support.

2.3. Boolean rings of projections in Banach spaces

Let E be a Banach space and $L(E, E)$ the algebra of all continuous linear operators on E (in general not commutative). The idempotents of the ring $L(E, E)$ are usually called projections. Thus, a Boolean ring of projections is to be understood as a Boolean ring of idempotents of the ring $L(E, E)$. This notion is equivalent to that used in the theory of self-adjoint and normal operators in Hilbert spaces and its generalization to Banach spaces. (cf. N. DUNFORD - J. SCHWARTZ [8], W. G. BADE [2]).

Every Boolean algebra can be realized as a Boolean algebra of projections in the Banach space $E = C(\mathrm{Spec}(\mathcal{B}))$.
The desired isomorphism is given by the correspondence

$$b \in \mathcal{B} \longmapsto P_b \in L(E, E), \text{ where } P_b(f) = \chi_{U(b)} \cdot f$$

for $f \in C(\mathrm{Spec}(\mathcal{B}))$, $\chi_{U(b)}$ being the indicator function of the clopen set $U(b)$ which corresponds to b.

From the preceding observation we get by a dual argument that the Boolean algebra $\mathcal{B}$ can be represented as a Boolean algebra of projections in the Banach space $M(\mathrm{Spec}(\mathcal{B}))$ of all regular bounded Borel measures on the representation space of $\mathcal{B}$. The adjoint operator for the projection P_b in $C(\mathrm{Spec}(\mathcal{B}))$ is the projection

$$P_b^*(\mu) = \chi_{U(b)} \cdot \mu \quad \text{for all} \quad \mu \in M(\mathrm{Spec}(\mathcal{B})).$$

2.4. Boolean algebra of all bands of an order complete vector lattice

Let V be an order complete vector lattice. A band B in V is defined to be an order convex linear subspace (= l-ideal in V) with the property that B is supremum closed in V. Bands are sometimes called closed l-ideals or in Russian literature (e.g. [28]) components of V. Bands are the kernels of full order homomorphisms (i.e. sup(inf) stable).

V satisfies the so-called Riesz decomposition theorem: For $A \subset V$ the set $A^{\perp} = \{b : b \in V \ |b| \wedge |a| = 0 \ \text{for all} \ a \in A\}$ is a band, $A^{\perp\perp}$ is the band generated by A and $V = A^{\perp\perp} \oplus A^{\perp}$. The system Band (V) of all bands of V ordered by inclusion forms a complete Boolean algebra, for which the lattice operations are the following:

$$(1) \quad B_1 \wedge B_2 = B_1 \cap B_2,$$

$$(2) \quad B_1 \vee B_2 = B_1 + B_2 \quad (= \mathrm{span}\,(B_1 \cup B_2)),$$

$$(3) \quad B' = B^{\perp},$$

$$(4) \bigwedge_{i \in \mathcal{J}} B_i = \bigcap_{i \in \mathcal{J}} B_i .$$

There is a 1-1 correspondence between the bands B of V and a special kind of projections $P : V \to V$, which are called full order projections and which are defined to be linear operators with the following properties:

(1) $P^2 = P$ and (2) $P(\sup A) = \sup P(A)$ for every bounded $A \subset V$.

This correspondence is established in the following way:

The projection P_B determined by the band B is the projection having B as range and annihilating its complement $B^{\perp}$.

$$B = \operatorname{Im} P_B , \quad B^{\perp} = \operatorname{Ker} P_B .$$

For every $x \in V$ $x \geqq 0$ it holds

$$P_B(x) = \sup \{ y : y \in B , \quad 0 \leqq y \leqq x \} \quad (*) .$$

Theorem 3

Let V be a complete vector lattice. The set $\mathcal{P}(V) \subset L(V, V)$ of all full order projections of V forms a complete Boolean algebra with respect to the operations

$$(1) \quad P \wedge Q = P \cdot Q = Q \cdot P ,$$

$$(2) \quad P' = I - P ,$$

$$(3) \quad P \vee Q = P + Q - P \cdot Q .$$

Further, it holds: The lattice order in $\mathcal{P}(V)$ is the same as the canonical projection order:

$$(4) \quad P \cdot Q = P \Leftrightarrow P \leqq Q \quad (\text{i.e. } Px \leqq Qx \text{ for all } x \geqq 0 \text{ in } V) \Leftrightarrow \operatorname{Im} P \subset \operatorname{Im} Q .$$

The Boolean algebra $\mathcal{P}(V)$ is isomorphic to the Boolean algebra Band (V) given by the correspondence $P \mapsto \operatorname{Im} P$.

Proof

First we show that all full order projections commute and $\mathcal{P}(V)$ is closed under multiplication. $P \in \mathcal{P}(V) \mapsto \operatorname{Im} P \in \text{Band}(V)$ is a 1-1 map and onto. It suffices to prove $P_{B \cap C}(x) = P_B(P_C(x))$ for all $B, C \in \text{Band}(V)$ and all $x \geqq 0$, which follows straightforward from $(*)$.

Furthermore, $\mathcal{P}(V)$ is closed with respect to $P \mapsto I - P$ for $P \in \mathcal{P}(V)$ since $I - P_B = P_{B^{\perp}}$. Thus $\mathcal{P}(V)$ is a Boolean ring of idempotents of the algebra $L(V, V)$. The Boolean operations are those described in Section 2.1.

Now we prove (4): The relation $P \leqq Q \Leftrightarrow \operatorname{Im} P \subset \operatorname{Im} Q$ can be seen easily.

The other part holds because $P_{B \cap C} = P_B \cdot P_C$. The remaining part of the theorem is clear. □

In the case when V is an order complete Banach lattice, the full order projections $P \in \mathcal{P}(V)$ are continuous, thus the Boolean algebra $\mathcal{P}(V)$ is a Boolean algebra of Banach space projections. The continuity of $P \in \mathcal{P}(V)$ follows from the relations $P|x| = |Px|$ and $P|x| \leqq |x|$. Then $\|Px\| \leqq \|x\|$ and $\|P\| \leqq 1$. Two types of Banach lattices are of special interest, namely the Kakutani M-spaces and the Kakutani L-spaces, because it is possible to identify the structure spaces of their Boolean algebras of bands.

Theorem 4

1. Let S be a Stone space. Then the Boolean algebra Band $(C(S))$ is isomorphic to the dual algebra of X. The isomorphism is given by the correspondence

$$\text{Band } (C(S)) \ni B \mapsto \text{supp } B \quad (= \text{cl}(\bigcup \text{supp } f ; f \in B) .$$

2. Let X be a compact Hausdorff space. Then the Boolean algebra Band $(M(X))$ is isomorphic to the dual algebra of the hyperstonian cover hX of X. The isomorphism is given by the correspondence

$$\text{Band } (M(X)) \ni B \mapsto \text{supp } hB \quad (=\text{cl}(\bigcup \text{supp } \mu^h, \mu \in B)) .$$

Proof

1. S being a Stone space is equivalent to $C(S)$ being a complete M-space. For every $f \in C(S)$, supp f $(=\text{cl}\{x : f(x) \neq 0\})$ is a clopen set. The equivalence of $|f| \wedge |g| = 0$ and $\text{supp } f \cap \text{supp } g = \emptyset$ is easy to verify. Thus, for a band $B \in C(S)$ the complement $B^{\perp}$ consists of all $f \in C(S)$ with $\text{supp } f \subset S \setminus \text{sup } B$. $B = B^{\perp\perp}$ implies $B = \{g : g \in C(S) , \text{supp } g \subset \text{supp } B\}$. Every clopen set $U \subset S$ determines also a band $B = \{f : \text{supp } f \subset U\}$. Therefore the ordered set Band $(C(S))$ is isomorphic to the ordered set $\mathbb{D}(S)$.

2. Let $N(hX)$ be the space of all hyperdiffuse measures on hX. In 1.2. we noticed that $M(X) \cong N(hX)$. For every $\mu \in M(X)$ let μ^h be the corresponding hyperdiffuse measure on hX. Every μ^h on hX gives a completely additive measure on the dual algebra $\mathbb{D}(hX)$ of all clopen sets of hX and vice versa (cf. [12]). Now for every σ-Boolean algebra $\mathbb{B}$ and for arbitrary positive σ-measures λ, ν on $\mathbb{B}$ it holds

$$\nu \in \{\lambda\}^{\perp\perp} \Leftrightarrow \nu \ll \lambda \quad (\text{i.e. } \lambda(b) = 0 \Rightarrow \nu(b) = 0,\ b \in \mathbb{B})$$

([23, p. 328]).

Thus $\mu, \nu \in M(X)$ implies: $|\mu| \wedge |\nu| = 0 \Leftrightarrow \text{supp } \mu^h \cap \text{supp } \nu^h = \emptyset$. $\square$

3. Compatible topologies on Boolean rings

3.1. Locally solid topologies

From topological algebra it is well known how to define the notion of a topological ring (see BOURBAKI [3]). Let $(R,+,\cdot)$ be a ring. A topology $\mathcal{O}$ on R is compatible with the algebraic structure if $(R,+,\mathcal{O})$ is a topological group and if the map $(a, b)\mapsto ab$ from $(R,\mathcal{O})\times(R,\mathcal{O})$ into $(R,\mathcal{O})$ is continuous. A ring endowed with a compatible topology is called a topological ring.

Using neighborhoods this can be described as follows: $(R,+,\cdot,\mathcal{O})$ is a topological ring iff the following conditions are satisfied: (N_1) For arbitrary points $x, y \in R$ and an arbitrary neighborhood W of $x - y$ there exist neighborhoods U of x and V of y with $U - V \subset W$. (N_2) For arbitrary points $x, y \in R$ and arbitrary neighborhood W of xy there exist neighborhoods U of x and V of y with $UV \subset W$. Due to the homogeneity of a topological ring the compatibility may be expressed by the neighborhood filter of the origin only. For a ring $(R, +, \cdot)$ to have a compatible topology $\mathcal{O}$ with a given filter base $\mathcal{B}$ on R as a neighborhood base of 0 with respect to $\mathcal{O}$ it is necessary and sufficient that the following conditions be satisfied:

NBZ 1. $0 \in B$ for all $B \in \mathcal{B}$.

NBZ 2. For every $B \in \mathcal{B}$ there is a $C \in \mathcal{B}$ such that $C + C \subset B$.

NBZ 3. For every $B \in \mathcal{B}$ there is a $C \in \mathcal{B}$ such that $-C \subset B$.

NBZ 4. For every $B \in \mathcal{B}$ and given $x \in R$ there is a $C \in \mathcal{B}$ with $xC \subset B$ and $Cx \subset B$.

NBZ 5. For every $B \in \mathcal{B}$ there is a $C \in \mathcal{B}$ such that $CC \subset B$.

Proposition

Let R be a Boolean ring and L the corresponding Boolean lattice. A topology $\mathcal{O}$ is compatible on R iff $\mathcal{O}$ is compatible with the lattice structure, i.e. $\mathcal{O}$ makes the maps $(x,y)\mapsto x\vee y$, $(x,y)\mapsto x\wedge y$ and $(x,y)\mapsto x - x\wedge y$ continuous.

Proof

This follows easily by observing that $x \vee y = x + y + xy$, $x \wedge y = xy$, $x - x\wedge y = x + xy$, $x + y = x \vee y - x \wedge y = (x - x\wedge y) \vee (y - x\wedge y)$. □

A topology $\mathcal{O}$ on a Boolean ring R is said to be locally solid (or locally order convex) iff every point has a neighborhood base of solid

(= order convex) sets

$$U: x,\ y \in U: x \leqq y \Rightarrow [x,\ y] = \{ z : x \leqq z \leqq y \} \subset U .$$

Of course, a set $U \subset R$ containing 0 is solid iff $UR = U$ (the inclusion $U \subset UR$ for every non-void $U \subset R$ holds). It is easily obtained that a topology on R is locally solid at the origin iff the family of maps $\{ \varphi_a : \varphi_a(x) := a \cdot x ,\ a \in R \}$ is equicontinuous at the origin.

Remark

Compatible topologies on Boolean rings with a solid neighborhood base of the origin are called uniformly compatible by J. L. COOPER [4],[5]. D. A. VLADIMIROV [27] used the notation monotone topology for such topologies while he called compatible topologies on Boolean algebras uniform topologies.

Theorem 5

Let X be a locally compact Boolean space and $C_{oo}(X)$ the lattice ring of all continuous real-valued functions with compact support on X. The locally solid compatible topologies on the Boolean ring I of idempotents of $C_{oo}(X)$ are exactly the trace topologies of all locally solid lattice algebra topologies on $C_{oo}(X)$.

Proof

Let $\mathcal{O}$ be a locally solid topology on $C_{oo}(X)$ with respect to which addition and multiplication in $C_{oo}(X)$ are continuous. Because the Boolean operations in I are related to the operations in $C_{oo}(X)$ by

$$f \boxplus g = f + g - 2fg \ , \ f \boxdot g = f \cdot g$$

the trace topology on I is compatible. The I-trace of every solid set $U \subset C_{oo}(X)$ is solid in I .

Now let conversely a compatible topology $\mathcal{P}$ on I be given which is solid at the origin. We shall construct a locally solid lattice ring topology $\mathcal{O}$ on $C_{oo}(X)$ with the I-trace $\mathcal{P}$. Let U be a $\mathcal{P}$ -neighborhood of 0 and $\varepsilon > 0$. We define $\widetilde{U_\varepsilon} := \{ f : f \in C_{oo}(X)$, there is an $e \in U$ such that $(1-e) \cdot |f| < \varepsilon \}$.
Then the following conditions are satisfied:

(1) $\widetilde{U_\varepsilon} \cap \widetilde{V_\varepsilon} \supset \widetilde{(U \cap V)_\varepsilon}$.
(2) $\widetilde{U_\varepsilon} \subset \widetilde{U_\delta}$ for $\varepsilon < \delta$.
(3) $\alpha \widetilde{U_\varepsilon} \subset \widetilde{U_\varepsilon}$ for all reals α with $|\alpha| < 1$.
(4) For every $f \in C_{oo}(X)$ there exists an $\alpha \in \mathbb{R}$ such that $f \in \alpha \widetilde{U_\varepsilon}$.
(5) $\widetilde{U_\varepsilon} + \widetilde{U_\varepsilon} \subset \widetilde{(U \vee U)_{2\varepsilon}}$.

(6) $\widehat{U}_\varepsilon \cdot \widehat{U}_\varepsilon \subset \widehat{(U \vee U)}_{\varepsilon^2}$.

(7) Every $\widehat{U}_\varepsilon$ is absolutely order convex .

(8) For every $\varepsilon < 1$ $\widehat{U}_\varepsilon \cap I = U \cdot I$.

Thus the set $\{ \widehat{U}_\varepsilon : U$ $\mathcal{P}$-neighborhood of 0 in I , $\varepsilon > 0 \}$ forms a solid filter base of the origin in $C_{oo}(X)$ and this is the neighborhood filter base of a unique lattice algebra topology $\mathcal{O}$ on $C_{oo}(X)$. According to (8) the I-trace of $\mathcal{O}$ coincides with $\mathcal{P}$ for a topology $\mathcal{P}$ which is locally solid at the origin. □

Corollary

Every compatible Boolean ring topology which is locally solid at the origin is locally solid at each point.

Proof

This is no trivial statement because order and addition in a Boolean ring are not compatible. Nevertheless, the proof of the statement follows the lines of the preceding proof. The constructed topology for $C_{oo}(X)$ is locally solid at every point because order and addition are compatible in $C_{oo}(X)$. □

The following three extremal topologies for the algebra lattice $C_{oo}(X)$ offer themselves.

1. Norm topology 2. compact open topology and 3. simple topology.

By the norm topology in $C_{oo}(X)$ the discrete topology on the Boolean ring I of idempotents is induced. (The trace of every ε-ball in $C_{oo}(X)$ for $0 < \varepsilon < 1$ yields only the zero-element in I .)

Proposition

The compact open topology in $C_{oo}(X)$ generates a locally solid compatible Boolean ring topology on the Boolean ring I of idempotents. By interpretation of I as the ring of compact-open sets this topology on I has the closed-set topology $\mathcal{T}_K$ induced by all compact sets in the sense of the upper semifinite topology as an open neighborhood base at the origin. The uniform completion of the topological Boolean ring I yields the Boolean ring of all idempotents of the ring $C(X)$ of all continuous functions on X .

Proof

For topologies in the hyperspace of closed sets of a topological space X we refer to [18] and [11] . In [11] we considered the topology $\mathcal{T}_K$ for $F(X) = \{ F : F$ closed subset of $X \}$. $\mathcal{T}_K$ possesses a base given by the sets $\langle K \rangle = \{F : F \cap K = \emptyset , F \in F(X)\}$, K compact. The compact open topology on $C_{oo}(X)$ yields the following open neighborhood base for I at the origin:

$$U(K) = \{ u : u \in I , \text{supp } u \cap K = \emptyset \}, \quad K \text{ compact.}$$

The Boolean ring $\tilde{I}$ of all idempotents of $C(X)$ is complete with respect to the compact open topology and I is dense in $\tilde{I}$. □

Proposition

The simple topology in $C_{oo}(X)$ induces a locally solid compatible Boolean ring topology on the Boolean ring I of idempotents. This topology is a precompact ring topology on I . The uniform completion of this topological Boolean ring I yields the so-called perfect completion of I .

Proof

Following F. B. WRIGHT [29] , the embedding of the Boolean ring I into the Boolean ring $\tilde{I}$ of all idempotents of $\mathbb{R}^X$ will be called the perfect completion. $\tilde{I}$ is compact with respect to the simple topology as the Tichonov product theorem shows. I is dense in $\tilde{I}$, since every indicator function of a single point is the pointwise limit of all elements of I whose support contains the point. Then every indicator function of an arbitrary non-void set S of X is the pointwise limit of all indicator functions with finite support contained in S . Thus the simple topology on I is a precompact ring topology which yields the perfect completion. □

3.2. The Riesz topology for functions and the induced Boolean ring topologies

What can be said about the possibility to get other algebra lattice topologies in $C_{oo}(X)$ for arbitrary locally compact spaces X ? Via the order dual of the Riesz space $C_{oo}(X)$ we have the possibility to introduce algebra lattice topologies. The order dual of $C_{oo}(X)$ is the vector lattice $M(X)$ of all Radon measures on X . Let $\mu \in M(X)$. We consider the topology of convergence in measure for $C_{oo}(X)$. This topology in $C_{oo}(X)$ will be called the Riesz topology induced by μ , because the notion of convergence in measure goes back to F. RIESZ [22]. If we start with a family $F \subset M(X)$ the supremum of all Riesz topologies induced by $\mu \in F$ will be called the Riesz topology induced by F . Let us recall some facts about the Riesz topology. For every $\varepsilon > 0$ let

$$U_\varepsilon^\mu := \{f : f \in C_{oo}(X) \text{ with } |\mu|(\frac{|f|}{1+|f|}) < \varepsilon\}.$$

Then the following conditions are fulfilled.

(1) $U_\varepsilon^\mu \subset U_\delta^\mu$ for $0 < \varepsilon < \delta$,

(2) $\alpha U_\varepsilon^\mu \subset U_\varepsilon^\mu$ for all $|\alpha| < 1$,

(3) for every $f \in C_{oo}(X)$ there exists an $\alpha \in \mathbb{R}$ such that $\alpha f \in U_\varepsilon^\mu$,

(4) $U_\varepsilon^\mu + U_\varepsilon^\mu \subset U_{2\varepsilon}^\mu$,

(5) $U_\varepsilon^\mu \cdot U_\varepsilon^\mu \subset U_{2\varepsilon}^\mu$.

(6) Every U_ε^μ is absolute order convex.

Thus the Riesz topology on $C_{oo}(X)$ induced by the Radon measure μ is an algebra lattice topology which is pseudo-metrizable. The trace topology on the Boolean ring I of idempotents of $C_{oo}(X)$ is precisely the Nikodym topology on I induced by μ. For a Boolean ring R which admits a positive measure $\mu > 0$, NIKODYM [20],[21] introduced a pseudo-metric $\rho(a, b) = \mu((a \vee b) - (a \wedge b))$. If we start with a $\mu \in M(X)$, we get for idempotents φ, ψ of $C_{oo}(X)$

$$|\mu|\left(\frac{|\varphi - \psi|}{1+|\varphi-\psi|}\right) = \frac{1}{2}|\mu|(|\varphi - \psi|) = \frac{1}{2}|\mu|((\varphi \vee \psi) - (\varphi \wedge \psi)).$$

Theorem 6

Let X be a compact Hausdorff space. Two families $F_1, F_2 \subset M(X)$ of Radon measures on X generate the same Riesz topologies on C(X) iff the generated bands $B(F_1)$, $B(F_2)$ in M(X) coincide. Thus the Riesz topologies on C(X) with respect to the inclusion form a Boolean algebra which is isomorphic to the algebra of bands of M(X).

Proof

Let $F \subset M(X)$ and let $\hat{F}$ be the l-ideal generated by F :

$$\hat{F} = \{\nu : \nu \in M(X) \quad |\nu| \leq \sum_{i=1}^{n} \alpha_i \cdot |\mu_i| \quad \alpha_i \in \mathbb{R}^+, \quad \mu_i \in F\}.$$

(1) We will show for the R-topologies: $\mathcal{O}(F) = \mathcal{O}(\hat{F})$.

(2) Then we will show $\mathcal{O}(cl(\hat{F})) = \mathcal{O}(\hat{F})$.

Because in every L-space the norm-closed l-ideals and the bands are the same objects we have by (1) and (2)

$$\mathcal{O}(F) = \mathcal{O}(B(F)).$$

(3) Then we prove for bands $B_1, B_2 \in$ Band (M(X)):

$$B_1 \subsetneqq B_2 \Rightarrow \mathcal{O}(B_1) \subsetneqq \mathcal{O}(B_2).$$

Ad (1): a) $U_\varepsilon^\mu \cap U_\varepsilon^\nu \subset U_{2\varepsilon}^{|\mu|+|\nu|}$,

b) $U_{\varepsilon}^{\alpha\mu} = U_{\frac{\varepsilon}{|\alpha|}}^{\mu} \quad \alpha \neq 0$,

c) $|\nu| \leqq |\mu| \Leftrightarrow U_1^{\mu} \subset U_1^{\nu}$.

In c) the implication from the left to the right is obvious.
Let $h \in C(X)$ with $0 \leqq h$ and $|\mu|(h) > 0$. Then for every $\alpha \in \mathbb{R}$ for which $0<\alpha<1$,

$$|\mu|\left(\frac{\alpha h}{|\mu|(h)}\right) < 1$$

thus $f \in U_1^{\mu}$ for $f \in C(X)$ with $\frac{\alpha h}{|\mu|(h)} = \frac{f}{1+f}$. By the assumption $f \in U_1^{\nu}$, i.e. $|\nu| \left(\frac{\alpha h}{|\mu|(h)} \right) < 1$. Therefore $|\nu|(h) \leqq |\mu|(h)$.
Ad (2): Let $\mu_n \in \hat{F}$ and $\|\mu_n - \mu\| \to 0$. Then $U_{\varepsilon}^{\mu_m} \subset U_{2\varepsilon}^{\mu}$ for $\|\mu_m - \mu\| < \varepsilon$, thus the topology $\mathcal{O}(\hat{F})$ is finer than the topology $\mathcal{O}(\mu)$.
Ad (3): It is clear that $\mathcal{O}(B_1) \subset \mathcal{O}(B_2)$ for $B_1 \subset B_2$. For $0 \leqq \nu \in B_2$ with $\nu \notin B_1$ we must prove $\mathcal{O}(\nu) \not\subset \mathcal{O}(B_1)$. In the case of $\mathcal{O}(\nu) \subset \mathcal{O}(B_1)$ we should have a measure $0 \leqq \mu \in B_1$ with $U_1^{\mu} \subset U_1^{\nu}$. Then by c) of (1) $0 \leqq \nu \leqq \mu$, which contradicts $\nu \notin B_1$. The proof is now complete since the bands of $M(X)$ form a Boolean algebra.□

Together with the theorem of Section 3.1 we get the following

Corollary

Let $\mathbb{B}$ be a Boolean algebra. Then the Nikodym topologies on $\mathbb{B}$ generated by families $F \subset M(\mathbb{B})$ of measures form with respect to inclusion a Boolean algebra which is isomorphic to the Boolean algebra of all bands of $M(\mathbb{B})$. The Nikodym topology on $\mathbb{B}$ generated by $F \subset M(\mathbb{B})$ is the trace of the Riesz topology in $C(\mathrm{Spec}(\mathbb{B}))$ induced by F .

3.3. The underlying Boolean algebras of topological semifields

In the theory of topological semifields founded by M. YA. ANTONOVSKI, V. G. BOLTJANSKII and T. A. SARYMSAKOV [1] complete Boolean algebras with compatible locally solid topologies are of special significance. Let us understand by an ABS-Boolean algebra such complete Boolean algebra $\mathbb{B}$ which is equipped with a Hausdorff compatible locally solid topology which is smaller than the order topology on $\mathbb{B}$. These and only these Boolean algebras are underlying Boolean algebras of topological semifields. If we start with a hyperstonian space X we get by the Nikodym topology $\mathcal{O}(N(X))$ (of all hyperdiffuse measures on X) on the dual algebra $\mathbb{D}(X)$ an ABS-topology. $C(X)$ with the Riesz topology $\mathcal{O}(N(X))$ will be a canonical associated topological semifield. For these results see a forthcomming paper of the author [14].

References

[1] M. YA. Antonovski, V. G. Boltjanskii and T. A. Sarymsakov: A survey on the theory of topologcal semifields. Uspehi Mat. Nauk 21 (1966), 185 - 218 . (Russian.)

[2] W. G. Bade: On Boolean algebras of projections and algebras of operators. Trans. Amer. Math. Soc. 80 (1955), 345 - 360.

[3] N. Bourbaki: Topologie générale. Paris 1951-58.

[4] J. L. Cooper: Convergence of families of completely additive set functions. Quart. J. Math. Oxford Ser. 20 (1949), 8 - 21.

[5] J. L. Cooper: Topologies in rings of sets. Proc. London Math. Soc. (2) 52 (1951), 220 - 239.

[6] R. P. Dilworth: The normal completion of the lattice of continuous functions. Trans. Amer. Math. Soc. 68 (1950), 427 - 438.

[7] J. Dixmier: Sur certains espaces considérés par M. H. Stone. Summa Brasil. Math. 2 (1951), 151 - 182.

[8] N. Dunford and J. T. Schwartz: Linear operators. Interscience Publ. 1958.

[9] D. A. Edwards: A class of topological Boolean algebras. Proc. London Math. Soc. (3) 13 (1963), 413 - 429.

[10] J. Flachsmeyer: Topologische Projektivräume. Math. Nachr. 26 (1963), 57 - 66.

[11] J. Flachsmeyer: Verschiedene Topologisierungen im Raum der abgeschlossenen Mengen. Math. Nachr. 26 (1963-64), 231 - 237.

[12] J. Flachsmeyer: Normal and category measures on topological spaces. Proc. Third Prague Topological Sympos. 1971, 109 - 116.

[13] J. Flachsmeyer: Dedekind-MacNeille extensions of Boolean algebras and of vector lattices of continuous functions and their structure spaces. General Topology and Appl. (To appear.)

[14] J. Flachsmeyer: Three examples of topological semifields. (To appear Russian.)

[15] A. Gleason: Projective topological spaces. Illinois J. Math. 2 (1958), 482 - 489.

[16] P. Halmos: Lectures on Boolean algebras. Van Nostrand Math. Studies 1967.

[17] S. Kaplan: On the second dual of the space of continuous functions. Trans. Amer. Math. Soc. 86 (1957), 70 - 90.

[18] E. Michael: Topologies on spaces of subsets. Trans. Amer. Math. Soc. 71 (1951), 152 - 182.

[19] H. Nakano: Über das System aller stetigen Funktionen auf einem topologischen Raum. Proc. Imp. Acad. Tokyo 17 (1941), 308 - 310.

[20] O. Nikodym: Sur une generalisation des integrals de M. J. Radon. Fund. Math. 15 (1930), 131 - 179.

[21] O. Nikodym: Sur les suites convergentes de fonctions parfaitement additives d'ensemble abstrait. Monatsh. Math. 40 (1933), 427 - 432.

[22] F. Riesz: Sur les suites de fonctions mesurables. C. R. Acad. Sci. Paris Sér. A - B 148 (1909), 1303 - 1305.

[23] Z. Semadeni: Banach spaces of continuous functions. PWN Warszawa 1971.

[24] R. Sikorski: Boolean algebras. Springer 1960.

[25] M. H. Stone: Applications of the theory of Boolean rings to general topology. Trans. Amer. Math. Soc. 41 (1937), 375 - 381.

[26] M. H. Stone: Boundedness properties in function lattices. Canad. J. Math. 1 (1949), 176 - 186.

[27] D. A. Vladimirov: Boolean algebras. Izdat. Nauka, Moscow 1969. (Russian.)

[28] B. Z. Vulich: Introduction to the theory of semi-ordered spaces. Goz. Izdat., Moscow 1961. (Russian.)

[29] F. B. Wright: Some remarks on Boolean duality. Portugal Math. 16 (1957), 109 - 117.

RECENT DEVELOPMENT OF THEORY OF UNIFORM SPACES

Zdeněk Frolík
Math. Inst. of the Czechoslovak Academy of Sci.
Žitná 25, 11000 Praha 1

Since the 4th Prague Symposium there has been a fast development of the theory of uniform spaces. Many old problems have been solved, and the solutions have opened new areas. Here I want to report just on two subjects: some sort of non-separable descriptive theory (cozero-sets, Baire sets etc.) and measure theory. My interest in uniform spaces originated in these two subjects. The preliminary introduction to uniform methods in descriptive theory was given in my talk on the 4th Prague Symposium and more definite program was introduced in my talks on conferences in Budva [F_2, 1972], Athens (Ohio 1972) and Pittsburgh [F_3, 1972]. A great help for me was the work of A.W. Hager; he was primarily interested in lattices of functions, however, he introduced to uniform spaces one of the most useful constructions (described in § 1 as $\mathcal{K}$ - c), and in the separable case observed that hereditarily metric-fine implies that the cozero sets form a σ-algebra. Also the work of F. Hansell on σ-discrete decomposability of disjoint completely Suslin-additive families in complete metric spaces (proved now by P. Holický and the present author for products of complete metric spaces by compact spaces) was basic for considering the program.

As concerns the measure theory on uniform spaces, introduced by D.A. Rajkov and independently by L.LeCam, one should consult [F_4],[F_5],[F_6] and recent papers by J. Pachl where one can find complete bibliography.

In 1973 I started a special seminar on uniform spaces which was attended from the very beginning by several graduate students who put a lot of will-power and enthusiasm in the work of the seminar. The results are published in seminar notes SUS 73-4, SUS 74-5 (very informal), and SUS 75-6 where one can find the details of all results presented here, and also references. In particular, SUS 73-4 contains a long survey of the material connected with cozero, Baire etc. sets and functions.

Uniform spaces theory could have always been attractive because of its conceptual value for topology, geometry and analysis. Unfortunately, difficult technical procedures were quite exceptional, and the answers to questions coming from the other fields seemed to be just formal reformulations which did not focuse on the problems. I feel that these days the theory of uniform spaces has its deep parts, and what is important, is giving the answers with "comments" which may be useful.

Let me just mention several groups of problems which are open for further investigation:

A. Uniform isomorphism problem for locally convex linear spaces: are any two uniformly isomorphic spaces isomorphic as topological linear spaces. It seems that no contra-example is known, however a number of results in positive direction has been given (Unflo, Mankiewicz, Vilímovský).

B. The complexity of uniform covers in general, in particular, the complexity of uniform covers of Banach spaces. The first non-trivial result is due to M. Zahradník, SUS 73-4, who showed that the uniform structure of no infinite-dimensional normed space is generated by ℓ_1-continuous partitions of unity. J. Pelant has developed a theory of combinatorial complexity of uniform covers using new technical tools, and the recent examples of spaces without point-finite basis due to V. Rödl and J. Pelant are very simple, however they are using more involved combinatorics (it should be mentioned that the first examples are due to Ščepin and Pelant).

C. Lattices and ordered sets of uniformities. Many results on atoms have been published by J. Pelant, J. Reiterman and P. Simon. In uniform spaces these topics are more complicated than in topological spaces, however ultrafilters play an important role.

D. The category of uniform spaces is quite interesting from the categorial point of view. Here I want just to mention that one is forced to use functors to be able to say something non-trivial and reasonable about various concrete problems; e.g. note the role of the plus and minus functors in what follows. Two classical facts are used frequently (Isbell, for the second also [Č]):

(1) every uniform space can be embedded into a product of complete metric spaces which can be chosen injective;

(2) every uniform space is a quotient of a "devil space" $D(\mathcal{F})$: $\mathcal{F}$ is a filter, say on a set X, the underlying set of $D(\mathcal{F})$ is

$Z \times X$, and the uniformity of $D(\mathcal{F})$ has the following covers $\mathcal{U}_F$, $F \in \mathcal{F}$, for a basis; $\mathcal{U}_F$ consists of all singletons $\langle 0,x \rangle$ and $\langle 1,x \rangle$ for $x \notin F$, and all two-point sets $\{\langle 0,x \rangle, \langle 1,x \rangle\}$ for $x \in F$.

It should be noted that the usage of the devil space was developed by M. Hušek.

E. Spaces of uniformly continuous mappings into topological linear spaces, in particular, into the reals. Two problems have been studied, existence of extensions and stability of the spaces with respect to the pointwise defined algebraic operations. In the real valued case J. Pelant, J. Vilímovský and the present author showed that the two problems are connected, and they described the largest coreflective class with the extension property. The general case has been studied by J. Vilímovský who showed that the situation is more complicated; the definite answers are not known. It should be noted that many results are implicitly contained in the literature on geometry of Banach spaces.

Now we are coming to the proper subject of my talk. The results will be stated in the terminology described in § 1.

§ 1. Notation and basic constructions.

We start with a description of three constructions which will be used frequently in describing various functors.

A. If $\mathcal{C}$ is a coreflective class of uniform spaces (closed under sums and quotients, and, f co rse, isomorphisms) then the class sub $\mathcal{C}$ consisting of subspaces of spaces in $\mathcal{C}$ is coreflective, and if c is the coreflection on $\mathcal{C}$ then subc stands for the coreflection on sub $\mathcal{C}$. Moreover, the functor subc is evaluated as follows: if $X \hookrightarrow Y$, and if Y is injective, then subc $X \hookrightarrow cY$. This general result is formulated by J. Vilímovský, the method was invented by J. Isbell who used it in the case of topologically fine spaces (the method was also used by M. Rice for metric-t_f spaces).

Remark. It is natural to ask the following question: Under what conditions the product of two spaces in a coreflective class $\mathcal{C}$ belongs to $\mathcal{C}$? The problem was opened by J. Isbell and Poljakov for the case of proximally fine spaces, and reconsidered by V. Kůrková by showing "yes" if one of the spaces is compact (or more generally, precompact). Her work was followed by nice results of M. Hušek, recently jointly with M. Rice, in general setting.

B. If $\mathcal{C}$ is a coreflective class then the class her $\mathcal{C}$ of all

X such that each subspace of X belongs to $\mathcal{C}$ does not need to be coreflective (e.g. take (A.D.) Alexandrov spaces for $\mathcal{C}$). However, if it is, then her c denotes the corresponding coreflection provided that c is the coreflection on $\mathcal{C}$.

C. $\mathcal{X}$ - c spaces. Let $\mathcal{X}$ be a class of spaces, and let c be a coreflection. The class $\mathcal{X}$ - c consists of all Y such that if $X \in \mathcal{X}$, and $f: Y \longrightarrow X$ is uniformly continuous, then so is $f: Y \longrightarrow cX$. This construction was introduced to uniform spaces by A. Hager.

The rest of this § is devoted to constructs associated with the concept of "refinement" introduced explicitly in [F_3] . If $\mathcal{H}$ is a category (always concrete) denote by $Set_{\mathcal{H}}$ the category on objects of $\mathcal{H}$ such that $Set_{\mathcal{H}}(X,Y)$ is the set of all mappings of X into Y. Here $\mathcal{H}$ is usually the category U of uniform spaces. A refinement of $\mathcal{H}$ is any category between $\mathcal{H}$ and $Set_{\mathcal{H}}$. The following sequence of refinements of U will be used frequently:

$$U \hookrightarrow \mathcal{D} \hookrightarrow p \hookrightarrow coz \hookrightarrow t \hookrightarrow Set_U.$$

Recall [F] that $\mathcal{D}(X,Y)$ is the set of all distal mappings from X into Y (the preimages of discrete collections are discrete), p(X,Y) is the set of all proximally continuous mappings from X into Y, coz (X,Y) is the set of all coz-mappings of X into Y (preimages of coz-sets are coz-sets, the coz-sets are preimages of open sets under the uniformly continuous functions), and t(X,Y) is the set of all continuous functions of X into Y.

It should be remarked that every concrete functor of $\mathcal{H}$ into any category defines a refinement, and every refinement is generated in this way.

Indeed, if $L: \mathcal{H} \longrightarrow \mathcal{L}$ is a concrete functor define a refinement $\mathcal{R}_L$ as follows:

$$\mathcal{R}_L(X,Y) = \mathcal{L}(LX,LY).$$

Usually we write L for $\mathcal{R}_L$. For example, t is the usual functor of U into topological spaces, which assigns to each uniform space the induced topological space. Similarly for the functor p of U into proximal spaces. The refinement coz may be defined by the functor coz into paved spaces: coz X is the set X endowed with the collection of all coz-sets in X. On the other hand, if $\mathcal{L}$ is a refinement of $\mathcal{H}$, then $\mathcal{L}$ is generated by the obvious functor of $\mathcal{H}$ into the category $\langle \mathcal{H} \rangle_{\mathcal{L}}$ defined as follows: the objects are the equivalence classes under the relation:X and Y are isomorphic in

$\mathcal{L}$ under the identity mapping of X onto Y; the equivalence class containing X is denoted by $\langle X \rangle_{\mathcal{L}}$.

Remark. In general, the approach to the refinements by means of functors into other categories has many advantages, e.g. simplicity. On the other hand, there are natural refinements "from life", e.g. the refinement in the § on measure theory.

Consider a refinement $U \hookrightarrow \mathcal{L}$. Denote by Inv $(\mathcal{L})$ (more precisely, Inv $(U \hookrightarrow \mathcal{L})$) the class of all concrete functors $F: U \longrightarrow U$ such that X and FX are isomorphic in $\mathcal{L}$ by the identity mapping; these functors are called $\mathcal{L}$-preserving. Denote by $\mathrm{Inv}_+(\mathcal{L})$ or $\mathrm{Inv}_-(\mathcal{L})$ the class of all positive or negative functors in $\mathrm{Inv}(\mathcal{L})$, accordingly. Recall that a functor F is called positive (negative) if the identity mapping $X \longrightarrow FX$ ($FX \longrightarrow X$) is uniformly continuous for each X. Note that concrete reflections are just the idempotent positive functors, and similarly for coreflections.

If there exists the coarsest functor in $\mathrm{Inv}_+(\mathcal{L})$, it is called the plus-functor of $\mathcal{L}$ and denoted by $\mathcal{L}_+$. Similarly, $\mathcal{L}_-$ is defined. We say that $\mathcal{L}_+$ is strong if $\mathcal{L}_+$ is the coarsest element in $\mathrm{Inv}(\mathcal{L})$. Self-evidently the plus and minus functors are idempotent.

Before discussing the properties of + and - functors, let me recall the concepts of fine and coarse objects. An object X is called $\mathcal{L}$-fine if

$$U(X,Y) = \mathcal{L}(X,Y)$$

for each Y. If the relation is satisfied for all X, then Y is called $\mathcal{L}$-coarse.

The class of all $\mathcal{L}$-fine objects is coreflective (easy), and the corresponding coreflection is denoted by $\mathcal{L}_f$. Similarly, the reflection on $\mathcal{L}$-coarse objects is denoted by $\mathcal{L}_c$. It is easy to show that the following three conditions are equivalent:

1. $\mathcal{L}_f = \mathcal{L}_-$.
2. $\mathcal{L}_f$ preserves $\mathcal{L}$ (i.e. $\mathcal{L}_f \in \mathrm{Inv}(\mathcal{L})$).
3. $\mathcal{L}_-$ generates $\mathcal{L}$ (i.e. $\mathcal{L}(X,Y) = U(\mathcal{L}_-X, \mathcal{L}_-Y)$).

If the three conditions are satisfied then $\mathcal{L}$ is called fine (this is equivalent to the statement that U is reflective in $\mathcal{L}$). Similarly for "coarse".

Classical simple results say that p is coarse, and t is fine. It is easy to show that $t_+ = p$ (strong), and it follows from the fact that $p(X \times X)$ uniquely determines X, that p_- is the identity (strong).

It is interesting that sub p_f is the identity (Hušek, Vilímovský), and in November 1976 J. Pelant (with certain help of P. Pták) showed that sub t_f is the locally fine coreflection, solving an old problem of J. Isbell.

It is shown in [F_3] that $\mathcal{D}$ is coarse. Note that $\mathcal{D}$ and related larger refinements have been studied by P. Pták.

The rest is devoted to several refinements related to descriptive theory and measure theory.

§ 2. Refinement coz .

Simple examples show that coz is neither coarse nor fine. It is easy to show that coz_c is the reflection on the indiscrete spaces. For many results on coz_f we refer to SUS 73-4, and SUS 74-5. Here I want just indicate the results on + and - functors. If $\mathcal{L}$ is a refinement, denote by $\mathcal{L}^2$ the refinement consisting of all $f: X \longrightarrow Y$ such that $f \times f: X \times X \longrightarrow Y \times Y$ is in $\mathcal{L}$.

Theorem. $coz_- = (coz^2)_f$ = metric - t_f. The functor coz_- is evaluated at X as follows:

a. The coz-sets in $X \times X$ containing the diagonal form a basis for the vicinities of the diagonal of $coz_- X$.

b. σ-discrete completely coz-additive covers of X form a basis for the uniform covers of $coz_- X$.

There is no reasonable description of the morphisms in coz_- except for the one in Theorem, the obvious conjectures fail to be true. It should be remarked that M. Rice found independently a description of metric-t_f similar to that in (b). The present author characterized metric-t_f spaces by several other properties (ℓ_∞ -partitions are ℓ_1, uniformly continuous maps into metric spaces are preserved by taking continuous limits). It seems that coz_- is one of the most useful functors.

It is obvious that $coz_- X = coz_f X$ iff $coz_- X$ is proximally fine. By general method (M. Rice, see A in Introduction), or directly one can show

$$\text{sub } coz_- = (\text{complete metric}) - t_f,$$

and it follows from the factorization theorem of G. Tashjian (coz-mappings of products into metric spaces factorize through countable sub-products) that

$$\text{sub coz}_- = \text{sub coz}_f.$$

Very useful is the coreflection her coz_ , which is called the measurable coreflection. A space X is in her coz_ iff it is in coz_ and coz X is a σ-algebra iff uniformly continuous mappings into metric spaces are closed under taking of pointwise limits of sequences.

It is easy to show that $coz_+ = p$, and it can be proved that $(coz_-)_+ = \mathcal{D}$.

§ 3. Refinement h coz .

The hyper-coz sets in a uniform space are the elements of the smallest collection of sets which contains all coz-sets, and it is closed under taking σ-discrete unions. The hyper-coz mappings are defined obviously. The properties of the resulting refinement h coz are similar to those of coz, however the proofs are more involved. Clearly

$$U \hookrightarrow h\ coz \hookrightarrow t,$$

and neither $coz \subset h\ coz$ nor $h\ coz \subset coz$.

Theorem. $h\ coz_- = (h\ coz^2)_f = coz_- \circ \lambda = coz_- \circ \text{sub}\ t_f$. In addition, h coz_ is evaluated at X as follows: the hyper-coz sets in X X containing the diagonal form a basis for the vicinities of the diagonal.

Corollary. $\text{sub}\ (h\ coz_-) = \text{sub}\ t_f$.

It follows from Theorem that to evaluate h coz_ at X it is enough to know $coz\ \lambda X = h\ coz\ X$, and $\mathcal{D}_c \circ \lambda X$. The distal structure of λX may be much finer than that of X. Therefore another functor is of certain interest, namely $(\text{metric} \times \text{compact}) - t_f$. This coreflection is evaluated as in (b) in Theorem in § 2 with coz replaced by h coz.

It can be proved that

$$h\ coz_+ = (h\ coz_-)_+ = \mathcal{D} .$$

§ 4. Refinements Ba and h Ba.

The Baire sets in X are the elements of the smallest σ-algebra containing the coz-sets. The hyper-Baire sets in X are the elements of the smallest σ-algebra containing the coz-sets, and closed under discrete unions. The set of all Baire-measurable mappings of X into Y, called simply Baire mappings, is denoted by Ba (X,Y). Similarly h Ba (X,Y) stands for the set of all hyper-Baire mappings of X into Y. The properties of the resulting refinements Ba and hBa depend on the model of set theory used, and the absolute results I know require quite deep properties of Suslin sets. Obviously

$$U \hookrightarrow \mathrm{coz} \hookrightarrow \mathrm{Ba} \hookrightarrow \mathrm{Set}_U,$$

$$U \hookrightarrow \mathrm{h\,coz} \hookrightarrow \mathrm{hBa} \hookrightarrow \mathrm{Set}_U,$$

and it is easy to see that Ba and hBa are unrelated. Also $\mathrm{Ba}_+ = p$, and $\mathrm{hBa}_+ = \mathcal{D}$. Certainly, none of the two refinements is coarse or fine.

It can be shown that Ba_- and hBa_- exist, however no description is known, and we shall see that the evaluation of the two functors at metric spaces depends on the model of set theory used. The absolute results are:

A. $\mathrm{Ba}_f X = \mathrm{Ba}_- X = \mathrm{her}\ \mathrm{coz}_- X$

if X is a complete metric space.

B. $\mathrm{hBa}_f X = \mathrm{hBa}_- X = \mathrm{her}\ (\mathrm{compact} \times \mathrm{complete\ metric} - t_f)$

if X is the product of a compact space by a complete metric space.

It is easy to show that A holds for all separable metric spaces under CH, and A does not hold for $Q \subset R$ (an uncountable subset of the reals such that each subset of Q is a G_δ , in particular, Ba Q is the power set of Q). J. Fleisner has announced a result which implies that A is consistent for all metric spaces of cardinal $\leq \omega_1$, and A holds for all metric spaces in a model.

For the proof of the absolute results one needs to know the following Lemma which was proved recently by P. Holický and the present author:

if X is a hyper-analytic space, in particular, if X is the product of a complete metric space by a compact space, then every disjoint completely Suslin-additive family in X is σ-discretely decomposable.

It should be remarked that for A one needs the case when X is complete metric, and this case is due to F. Hansell. Recall that $\{X_a\}$ is σ-discretely decomposable if there exists a family $\{X_{a,n}\}$ such that each $\{X_{a,n}\}_a$ is discrete, and $X_a = \bigcup \{X_{a,n} \mid n \in \omega\}$.

For A one needs also a Lemma due to D. Preiss: every disjoint completely Baire-additive family is of a bounded class.

The statements of the categorial consequences are left to the of $\mathcal{M}_U(X)$ into E. Clearly $\mathcal{M}_U(X)$ is uniquely determined up to an isomorphism preserving σ^X. Clearly every $f \in U(X,Y)$ extends uniquely to a continuous linear map $\mathcal{M}_U(f)$ of $\mathcal{M}_U(X)$ into $\mathcal{M}_U(Y)$, thus $\mathcal{M}_U$ is a functor. The elements of $\mathcal{M}_U(X)$ are called uniform measures on X.

It is not difficult to show that $\mathcal{M}_U(X)$ can be identified with the set of all $\mu \in \mathcal{M}(X)$ which are continuous in the pointwise topology on each UEB set, endowed with the topology of uniform convergence on UEB sets. The embedding σ^X assigns to each $x \in X$ the evaluation at x, i.e. the Dirac measure at x. One can show that the set $\mathcal{M}ol(X)$ of all molecular measures, i.e. the linear space generated by Dirac measures, is dense in $\mathcal{M}_U(X)$. Thus $\mathcal{M}_U(X)$ is a completion of $\mathcal{M}ol(X)$ endowed by the topology of uniform convergence on UEB sets. The topology of $\mathcal{M}_U(X)$ is called the uniform topology. The linear space $U_b(X)$ is the dual of $\mathcal{M}_U(X)$, and what is important for our purposes, on the positive cone $\mathcal{M}_U^+(X)$ the uniform topology coincides with the weak topology (= $\sigma(\mathcal{M}_U(X), U_b(X))$).

The concept of uniform measure may be useful for measure theory because uniform measures are preserved by projective limits, σ-additive measures defined on σ-algebras and also cylindrical measures can be viewed as uniform measures. One can define vector valued uniform measures, and develop a nice theory of integration; the basic result for this purpose is a recent theorem of J. Pachl which says that relatively weakly compact subsets of $\mathcal{M}_U(X)$ are relatively compact. It is natural to ask whether some questions about uniform measures can be reduced to consideration of uniform measures on very simple uniform spaces. One problem of this sort is considered here.

What can be said about negative functors F of uniform spaces such that the middle vertical arrow in the diagram

$$\begin{array}{ccccc} X & \hookrightarrow & \mathcal{M}_U^+(X) & \hookrightarrow & \mathcal{M}_U(X) \\ \uparrow & & \uparrow & & \uparrow \\ FX & \hookrightarrow & \mathcal{M}_U^+(FX) & \hookrightarrow & \mathcal{M}_U(FX) \end{array}$$

is a homeomorphism. The answer is that there exists the finest one, and the refinement $\mathcal{M}$ of U generated by the finest one can be described as follows:

$f \in \mathcal{M}(X,Y)$, iff $\mathcal{M}ol^+(f)$ extends to a continuous mapping of $\mathcal{M}_U^+(X)$ into $\mathcal{M}_U^+(Y)$.

Of course, $\mathcal{Mol}^+(f)$ is the restriction to $\mathcal{Mol}^+(X)$ of the extension $\mathcal{Mol}(f)$ of f to a linear mapping of $\mathcal{Mol}(X)$ into $\mathcal{Mol}(Y)$.

I don't know any direct proof. In my proof one considers at the same time a functor by means of playing with "true" Radon measures, and working with both $\mathcal{M}$ and the functor constructed one finally shows the proposition about $\mathcal{M}$, and proves that the functor is $\mathcal{M}_f$. The construction of $\mathcal{M}_f$ is based on the following description of uniform measures [F_5] :

a measure $\mu \in \mathfrak{M}(X)$ is uniform iff $\check{\mu}$ is sitting on $K(\mathcal{U})$ as a Radon measure for each uniform cover $\mathcal{U}$ of X.

Here $K(\mathcal{U})$ is the union of closures in $\check{X}$ of the elements of $\mathcal{U}$.

Remark. This description implies that on a complete metric space the uniform measures are just the Radon measures.

The description of $\mathcal{M}_f$:

$\mathcal{M}_f$ is X endowed with the coarsest uniformity such that all the identity maps

$$\mathcal{M}_f X \longrightarrow t_f U$$

are uniformly continuous, where U runs over all $U \subset \check{X}$ such that for each uniform measure μ on X the measure $\check{\mu}$ is sitting on U as a Radon measure.

The spaces $\mathcal{M}_f X$ have good properties. They are locally fine (but $\mathcal{M}_f \neq \lambda$), and σ-additivity of all uniform measures on X is described by simple properties of X as follows.

Theorem. The following properties of X are equivalent:

1. Each uniform measure on X is σ-additive (i.e. $f_n \downarrow 0$, $f_n \in U_b(X) \Longrightarrow \mu(f_n) \longrightarrow 0$).

2. $\mathcal{M}_f X$ is metric - t_f (i.e. $coz_- \mathcal{M}_f X = \mathcal{M}_f X$).

3. $\mathcal{M}_f X$ is inversion-closed (i.e. if $f > 0$ is a uniformly continuous function then so is $1/f$).

4. If $f_n \downarrow 0$, and $\{f_n\} \subset U_b(\mathcal{M}_f X)$, then $\{f_n\}$ is UEB.

Remark. The properties 3. and 4. are always equivalent (M. Zahradník, SUS 73-4).

The details of § 6 will appear in SUS 76-7, see also [F_6] .

It is an open problem whether there exists a non-trivial positive functor F such that the middle vertical arrow is an analogous diagram is a homomorphism (even a bijection).

References.

Isbell J.R.: Uniform spaces, Math. Surveys (12), 1964.

Frolík Z.

[F_1] Topological methods in measure theory and the theory of measurable spaces, General topology and its relations to modern analysis and algebra III (Proc. 3rd Prague Symposium, 1971), Academia, Prague, 1972, 127-139.

[F_2] Interplay of measurable and uniform spaces, Proc. 2nd Symposium on topology and its applications (Budva 1972), Beograd 1973, 96-99.

[F_3] Basic refinements of the category of uniform spaces, Proc. 2nd Int. Topological Symp., Lecture notes in math. 378, Springer-Verlag 1974, 140-159.

[F_4] Mesures uniformes, C.R. Acad. Sci. Paris, t. 277, Série A, 105-8.

[F_5] Représentation de Riesz des mesures uniformes, C.R. Acad. Sci. Paris, t. 277, Série A, 163-166.

[F_6] Measure-fine uniform spaces, Measure Theory Oberwolfach 1975, Lecture notes in math., Springer-Verlag 1976, 403-13.

Publications of Math. Inst. of ČSAV in Prague:

[SUS 73-4] Seminar Uniform Spaces 1973-4, Prague 1975.

[SUS 74-5] Seminar Uniform Spaces 1974-5, Prague 1975 (very informal notes).

[SUS 75-6] Seminar Uniform Spaces 1975-6, Prague 1976.

STABILITY OF BANACH ALGEBRAS

B.E. Johnson
(Newcastle upon Tyne, England)

Given a mathematical structure, it is of interest mathematically to know whether other structures 'near' to it in some sense share various properties of the original structure. If the structure is used as part of a mathematical model of some physical situation then this question is particularly important as the parameters determining the system cannot be evaluated with complete accuracy. Thus it will not be known exactly which of a number of systems is involved and so properties which are common to all these systems have a special significance.

We shall consider a Banach algebra $\mathfrak{A}$ with multiplication π. Juxtaposition of elements of $\mathfrak{A}$ will denote their π product. We shall also consider other multiplications, that is continuous associative bilinear maps ρ; $\mathfrak{A} \times \mathfrak{A} \to \mathfrak{A}$. We assume $||ab|| \leq ||a||\,||b||$ but shall not assume $||\rho(a,b)|| \leq ||a||\,||b||$ $(a,b \in \mathfrak{A})$. The strongest possible result is that if ρ is sufficiently near π then $(\mathfrak{A},\pi)$ and $(\mathfrak{A},\rho)$ are isomorphic, that is there is a continuous linear bijection T; $\mathfrak{A} \to \mathfrak{A}$ with $T\rho(a,b) = (Ta)(Tb)$ $(a,b \in \mathfrak{A})$. This is equivalent to saying that ρ has the form

$$\rho(a,b) = T^{-1}((Ta)(Tb)) \qquad \text{(i).}$$

It is easy to see that defining ρ by formula (i) gives a multiplication on $\mathfrak{A}$ so our question is whether (i) is a necessary as well as a sufficient condition for ρ to be a multiplication. This question was first raised by J.L. Taylor and the result we give below was also obtained by Raeburn and Taylor [10]. As in [3] we write $\mathfrak{L}^n(\mathfrak{A})$ for the Banach space of continuous n linear functions with variables and values in $\mathfrak{A}$.

DEFINITION 1. A Banach algebra $\mathfrak{A}$ is <u>stable</u> if there exists $\varepsilon > 0$ such that for each multiplication ρ on $\mathfrak{A}$ with $||\rho-\pi|| < \varepsilon$ there is a non-singular element T of $\mathfrak{L}^1(\mathfrak{A})$ with $\rho(a,b) = T^{-1}(T(a)T(b))$ $(a,b \in \mathfrak{A})$. ($||\rho-\pi||$ denotes the $\mathfrak{L}^2(\mathfrak{A})$ norm.)

To motivate our result consider the related problem of representing the multiplications in a one parameter family $\{\rho_t\}$, where $\rho_0 = \pi$, in the form $\rho_t(a,b) = T_t^{-1}(T_t a)(T_t b)$ for a suitably chosen one parameter family T_t from $\mathfrak{L}^1(\mathfrak{A})$. If

$$\left(\frac{d\rho_t}{dt}\right)_{t=0} = S$$

then differentiating the associative law

$$\rho_t(a,\rho_t(b,c)) = \rho_t(\rho_t(a,b),c)$$

and putting $t = o$ gives

$$S(a,bc) + a\,S(b,c) = S(ab,c) + S(a,b)c$$

that is

$$a\,S(b,c) - S(ab,c) + S(a,bc) - S(a,b)c = 0 \qquad \text{(ii)}.$$

Putting $R = (\frac{dT_t}{dt})_{t=0}$, differentiating the relation $T_t\rho_t(a,b) = T_t a T_t b$ and putting $t = 0$ gives

$$a\,R(b) - R(ab) + R(a)b = S(a,b) \qquad \text{(iii)}.$$

Thus under suitable conditions of differentiability we must solve (iii) for R given that S satisfies (ii). For any $R \in \mathfrak{L}^1(\mathfrak{A})$ (resp. $S \in \mathfrak{L}^2(\mathfrak{A})$) the left hand side of (iii) (resp. (ii)) defines an element $\delta^2 R$ of $\mathfrak{L}^2(\mathfrak{A})$ (resp. $\delta^3 S$ of $\mathfrak{L}^3(\mathfrak{A})$). It is easy to see, by direct substitution, that $\delta^3\delta^2 R = 0$ so the question of solving (iii) given (ii) is just the question of whether $\ker \delta^3 = \operatorname{im} \delta^2$. This has been considered extensively (see [3] and elsewhere).

THEOREM 2. *Let* $\mathfrak{A}$ *be a Banach algebra with* $\ker \delta^3 = \operatorname{im}\delta^2$ *and* $\operatorname{im}\delta^3$ *closed in* $\mathfrak{L}^3(\mathfrak{A})$. *Then* $\mathfrak{A}$ *is stable*.

Proof. Let $\epsilon > 0$ and let ρ be a multiplication on $\mathfrak{A}$ with $\|\pi-\rho\| < \epsilon$. By the open mapping theorem and the hypotheses that $\operatorname{im}\delta^2 = \ker\delta^3$, a closed subspace of $\mathfrak{L}^2(\mathfrak{A})$, and $\operatorname{im}\delta^3$ is closed, there exist $K, L > 0$ such that if $S \in \ker\delta^3$ and $T \in \mathfrak{L}^2(\mathfrak{A})$ there exist $R \in \mathfrak{L}^1(\mathfrak{A})$ and $S' \in \mathfrak{L}^2(\mathfrak{A})$ with $\delta^2 R = S$, $\|R\| \le K\|S\|$, $\delta^3 S' = \delta^3 T$ and $\|S'\| \le L\|\delta^3 T\|$.

We have

$$\begin{aligned}
&(\pi-\rho)((\pi-\rho)(a,b),c) - (\pi-\rho)(a,(\pi-\rho)(b,c)) = \\
&(ab)c - a(bc) + \rho(\rho(a,b),c) - \rho(a,\rho(b,c)) \\
&- \rho(ab,c) - \rho(a,b)c + \rho(a,bc) + a\rho(b,c) \\
&= (\delta^3\rho)(a,b,c) = \delta^3(\rho-\pi)(a,b,c)
\end{aligned}$$

so that $\|\delta^3(\rho-\pi)\| \le 2\epsilon^2$. Thus there is $S \in \mathfrak{L}^2(\mathfrak{A})$ with $\|S\| \le 2L\epsilon^2$ and $\delta^3 S = \delta^3(\rho-\pi)$. As $\rho-\pi-S \in \ker\delta^3$ there is $R \in \mathfrak{L}^1(\mathfrak{A})$ with $\delta^2 R = \rho-\pi-S$ and $\|R\| \le K\|\rho-\pi-S\| \le K(\epsilon+2L\epsilon^2) = K\epsilon(1+2L\epsilon) < 2K\epsilon < \frac{1}{2}$ provided $\epsilon < \operatorname{Max}((2L)^{-1},(4K)^{-1})$. Thus I+R is regular in $\mathfrak{L}^1(\mathfrak{A})$. Putting $\rho'(a,b) = (I+R)^{-1}[(I+R)a(I+R)b]$ and expanding $\rho(a,b) - \rho'(a,b)$ in powers of R, the constant and first degree terms are

$$\begin{aligned}
\|\rho(a,b) - ab + R(ab) - aR(b) - R(a)b\| &= \|(\rho-\pi-\delta^2 R)(a,b)\| \\
&= \|S(a,b)\| \\
&\le 2L\epsilon^2\|a\|\|b\|
\end{aligned}$$

whereas the other terms give

$$\begin{aligned}
&\Big\|\sum_{n=0}^{\infty}(-1)^n\, R^{n+2}(ab) - R^{n+1}(aRb) - R^{n+1}((Ra)b) + R^n(RaRb)\Big\| \\
&\le \sum_{n=0}^{\infty} 4\|R\|^{n+2}\|a\|\|b\| \\
&\le 4\|R\|^2(1-\|R\|)^{-1}\|a\|\|b\|
\end{aligned}$$

$$\leq 32K^2\varepsilon^2\|a\|\|b\| .$$

Thus $\|\rho-\rho'\| \leq (32K^2+2L)\varepsilon^2$. Now put $\rho_1(a,b) = (I+R)\rho((I+R)^{-1}a, (I+R)^{-1}b)$. We have

$$\|(\rho_1-\pi)(a,b)\| = \|(I+R)(\rho-\rho')[(I+R)^{-1}a, (I+R)^{-1}b]\|$$

so that

$$\begin{aligned}\|\rho_1-\pi\| &\leq \|I+R\|\|\rho-\rho'\|\|(I+R)^{-1}\|^2 \\ &\leq 8(32K^2+2L)\varepsilon^2 \\ &= M\varepsilon^2\end{aligned}$$

say. As ρ_1 is a multiplication we can apply the above argument with ρ replaced by ρ_1 obtaining R_1, S_1, ρ_2 and so on inductively. At the nth stage we have $\|\rho_n-\pi\| \leq M^{2^n-1}\varepsilon^{2^n}$. Thus if $\varepsilon < M^{-1}$ we have $\rho_n \to \pi$ as $n \to \infty$. Putting $W_n = (I+R_n)..(I+R_1)(I+R)$ we have, since $\|R_n\| < 2KM^{2^n-1}\varepsilon^{2^n}$, $W = \lim W_n$ exists and is regular because $\|W_n^{-1}\| \leq [(1-\|R\|)(1-\|R_1\|)\ldots(1-\|R_n\|)]^{-1}$ where the infinite product $\prod(1-\|R_i\|)$ is convergent to a non-zero limit. Thus

$$ab = \lim_n \rho_n(a,b) = \lim_n W_n\rho(W_n^{-1}a, W_n^{-1}b) = W\rho(W^{-1}a, W^{-1}b) .$$

Replacing a,b by Wa, Wb we get $\rho(a,b) = W^{-1}(Wa)(Wb)$.

The following Banach algebras are stable

(i) The algebra $C(X)$ of all bounded continuous complex valued functions on a topological space X.

(ii) The algebra $\mathcal{L}(\mathfrak{X})$ of all bounded linear operators on a Banach space $\mathfrak{X}$.

(iii) The algebra $\mathcal{LC}(\mathfrak{X})$ of all compact operators on a Banach space $\mathfrak{X}$ with an unconditional basis.

(iv) The algebra C_1 of trace class operators on a separable Hilbert space.

(v) The sequence algebras ℓ^1 and ℓ^∞ with pointwise multiplication.

(vi) Any type I or hyperfinite von Neumann algebra.

(vii) The group algebras $L^1(G)$, $M(G)$ of an amenable locally compact group G.

By contrast the following Banach algebras are not stable

(i)′ The algebra C_2 of Hilbert-Schmidt operators and, more generally, the von Neumann-Schatten classes C_p [1;p.1089] for $1 < p < \infty$.

(ii)′ The sequence algebra ℓ^p with pointwise multiplication for $1 < p < \infty$.

(iii)′ The group algebra $\ell^1(F_2)$ of the free group on 2 generators.

The positive results, with the exception of (i) (see Theorem 3 below) are all proved by showing that the hypotheses of Theorem 2 are satisfied. For (ii) this is a result of Kaliman and Selivanov [8], for (iii) the proof is a minor adaptation of the Hilbert space case [4; p.697], (vi) depends on results of Kadison and Ringrose ([6; Theorem 4.4] and [7; Theorem 3.1]) and (vii) follows from [3; Theorem 2.5 and Proposition 1.9] and [4; Theorem 4.4 and Example 4.2]. (iv) and (v) appear in [5;

§3B and 3C]. The counterexamples also appear in [5].

THEOREM 3. Let X be a topological space. Then C(X) is stable.

Proof. Since C(X) is a commutative C*-algebra it is (isometrically isomorphic with) $C(\Omega)$ for some compact Hansdorff space Ω. Thus the fact that $\ker \delta^3 = \mathrm{im}\, \delta^2$ is a result of Kamowitz [9; Theorem 4.7]. To show that the other hypothesis is satisfied we consider first the case in which X is a metric space with metric d. We shall show $\mathrm{im}\, \delta^3$ is closed by showing $\mathrm{im}\, \delta^3 = \ker \delta^4$ where $\delta^4;\ \mathcal{L}^3(\mathfrak{A}) \to \mathcal{L}^4(\mathfrak{A})$ is defined by

$$(\delta^4 T)(a,b,c,d) = aT(b,c,d) - T(ab,c,d) + T(a,bc,d) - T(a,b,cd) + T(a,b,c)d .$$

For each positive integer k define $f_k;\ [0,\infty) \to [0,1]$ by

$$f_k(t) = 0 \qquad 0 \le t \le 2^{-k-1} \qquad \text{or } 2^{-k+1} \le t$$

$$f_k(2^{-k}) = 1$$

and f_k is linear on $[2^{-k-1},2^{-k}]$ and $[2^{-k},2^{-k+1}]$. We define f_0 in the same way on $[0,1]$ and put $f_0(t) = 1$ for $t \ge 1$. For $x \in X$ let $g_{x,k}(y) = (f_k \circ d(x,y))^{\frac{1}{2}}$. As in Helemskii [2; §4] if we put $J_x(a) = \sum_k g_{x,k} \otimes g_{x,k}\, a$ then J_x is a map $M_x (= \{a; a \in C(X),\ a(x) = 0\}) \to M_x \hat{\otimes} M_x$ of norm at most 2. If $T \in \ker \delta^4$ put

$$S(a,b)(x) = -T(J_x(a-a(x)1),\ b - b(x)1)(x)$$
$$+\ a(x)T(1,1,b)(x) - b(x)T(a-a(x)1,1,1)(x)$$

where we have also used T to denote the map $M_x \hat{\otimes} M_x \times C(X)$ derived from T. To see that for each a,b, $S(a,b) \in C(X)$ it is enough to show that $x \to J_x(a-a(x)1)$ is continuous $X \to C(X) \hat{\otimes} C(X)$ for each $a \in C(X)$. Let $y \in X$. Since $J_x(a-a(x)1)$ is linear in a and $\|J_x(a-a(x)1)\| \le 4\|a\|$ it is enough to prove continuity at y for a in a dense subspace of C(X). Accordingly we suppose that a is constant in a neighbourhood $\{z;\ d(z,y) \le 2^{-n}\}$ of y. If $d(y,z) < 2^{-n-1}$ then $(a - a(z)1)(w) = 0$ for $d(z,w) < 2^{-n-1}$ so for such z

$$J_z(a-a(z)1) = \sum_{k=0}^{n+2} g_{z,k} \otimes g_{z,k}(a-a(z)1)$$

and the map $x \to g_{x,k}$ is continuous $X \to C(X)$ for each k.

Finally we show $\delta^3 S(a,b,c)(x) = T(a,b,c)(x)$. First consider the case $a,c \in M_x$. Then

$$\delta^3 S(a,b,c)(x) = -S(ab,c)(x) + S(a,bc)(x)$$
$$= T(J_x(ab),c)(x) - T(J_x(a),bc)(x)$$
$$= T(a,b,c)(x)$$

because

$$0 = \delta^4 T(g_{x,k}, g_{x,k}a,b,c)(x)$$

$$= -T(g_{x,k}^2 a,b,c)(x) + T(g_{x,k}, g_{x,k}ab,c)(x)$$

$$-T(g_{x,k}, g_{x,k}a, bc)(x)$$

so that summing over k and using $\sum_k g_{x,k}(y)^2 = 1$ $(y \neq x)$ we obtain the required relationship.

Now suppose $a = 1$. Then

$$\begin{aligned}\delta^3 S(1,b,c)(x) &= S(1,bc)(x) - S(1,b)(x)\ c(x)\\ &= T(1,1,bc)(x) - T(1,1,b)(x)\ c(x)\\ &= T(1,b,c)(x)\end{aligned}$$

by considering the identity $(\delta^4 T)(1,1,b,c) = 0$.

Last consider the case $a \in M_x$, $c = 1$.

$$\begin{aligned}\delta^3 S(a,b,c)(x) &= -S(ab,1)(x)\\ &= T(ab,1,1)(x)\\ &= T(a,b,1)(x)\end{aligned}$$

by considering the equation $\delta^4 T\ (a,b,1,1)(x) = 0$.

To treat the case of general X rather more attention to the constants K, L, M of Theorem 2 is necessary. By the proof of Kamovitz' theorem that $\ker \delta^3 = \operatorname{im} \delta^2$ given in [3; Proposition 8.2] we see $K = 1$. From the definition of S in terms of T above we see $\|S\| \leq 11\|T\|$ so for metric X we can take $L = 11$ giving $M = 432$. Thus, using the notation of Theorem 2 applied to this case

$$\begin{aligned}\|W_n - I\| &\leq (1+\|R\|)(1+\|R_1\|)\ \ldots\ (1+\|R_n\|) - 1\\ &\leq (1+2\varepsilon)(1+2M\varepsilon^2)\ \ldots\ (1+2M^{2^{n-1}}\varepsilon^{2^n}) - 1\\ &\leq \frac{M\varepsilon}{1-M\varepsilon}\end{aligned}$$

so $\|W-I\| < M\varepsilon(1-M\varepsilon)^{-1}$. From this we see $\|W^{-1}-I\| \leq 2M\varepsilon(1-M\varepsilon)^{-1}$. If $\varepsilon < (3M)^{-1}$ we get $\|W-I\| < \frac{1}{2}$ and $\|W^{-1}-I\| < 1$.

For each finite subset F of C(X) let $\mathfrak{A}_F$ be the smallest closed unital * subalgebra of C(X) containing F which is ρ closed. $\mathfrak{A}_F$ is separable as it is the closed linear span of elements which can be expressed by a finite number of applications of $\pi, \rho, *$ to elements of F and these form a denumerable set. Hence $\mathfrak{A}_F$ is isomorphic with $C(\Omega)$ for a compact metric space Ω. Thus by the theorem for metric X, if ρ is a multiplication on C(X) with $\|\rho-\pi\| < 1/1296$ then for each finite $F \subseteq C(X)$ there is an invertable element W_F of $\mathcal{L}^1(\mathfrak{A}_F)$ with $\rho(a,b) = W_F^{-1}(W_F(a)W_F(b))$ $(a,b \in \mathfrak{A}_F)$ and $\|W_F - I_{\mathfrak{A}_F}\| < \frac{1}{2}$, $\|W_F^{-1} - I_{\mathfrak{A}_F}\| < 1$. Suppose F,G are two finite subsets of C(X) with $\mathfrak{A}_F \subseteq \mathfrak{A}_G$ so that for $a,b \in \mathfrak{A}_F$, $W_F^{-1}(W_F a W_F b) = W_G^{-1}(W_G a W_G b)$, that is $\alpha = W_G W_F^{-1}$ is an isomorphism of $\mathfrak{A}_F$ into $\mathfrak{A}_G$ with $\|\alpha - \iota\| < \frac{1}{2}$ where ι is the injection $\mathfrak{A}_F \subseteq \mathfrak{A}_G$. If φ is a multiplicative linear functional on $\mathfrak{A}_G$ then $\alpha^*\varphi$ and $\iota^*\varphi$ are multiplicative linear functionals on $\mathfrak{A}_F$ with $\|\alpha^*\varphi - \iota^*\varphi\| < 2$ so $\alpha^*\varphi = \iota^*\varphi$. Hence $\varphi(\alpha a) = \varphi(a)$ for all $a \in \mathfrak{A}_F$ and all φ. This implies that $a = \alpha a$ and hence $W_G a = W_F a$ for all a in $\mathfrak{A}_F$. The map W on C(X) defined by

$$Wa = W_F a \qquad \text{if } a \in \mathfrak{A}_F$$

is thus well defined, bounded, linear and has $\|W-I_{C(X)}\| \leq \frac{1}{2}$ so that it is invertible. On each $\mathfrak{A}_F$, $W^{-1}a = W_F^{-1}a$ so that $\rho(a,b) = W^{-1}(WaWb)$ $a,b \in C(X)$.

REFERENCES

1. Dunford, N. and Schwartz, J.T., Linear operators, part 2, Interscience, New York, 1963.

2. Helemskii, A.Ja., On the homological dimension of normed modules over Banach algebras, (Russian)Mat. Sbornik, vol.81 (123) (1970), pp430-444, (English translation) Math. U.S.S.R. Sbornik, vol.10 (1970) pp399-411.

3. Johnson, B.E., Cohomology in Banach algebras, Memoirs of the American Mathematical Society, vol.127 (1972).

4. Johnson, B.E., Approximate diagonals and cohomology of certain annihilator Banach algebras, American J. Math., vol.94 (1972), pp685-698.

5. Johnson, B.E., Perturbations of Banach Algebras. To appear in Proc. London Math. Soc.

6. Kadison, R.V. and Ringrose, J.R., Cohomology of operator algebras I, Type I von Neumann algebras, Acta Math., 126 (1971), 227-243.

7. Kadison, R.V. and Ringrose, J.R., Cohomology of operator algebras II, Extended cobounding and the hyperfinite case, Ark. Mat. 9 (1971), 55-63.

8. Kaliman, Sh.I. and Selivanov, Yu.V., On cohomologies of operator algebras, (Russian), Vestnik Moskov Univ. Ser 1, Mat. Meh, 5 (1974), 24-27.

9. Kamowitz, H., Cohomology groups of commutative Banach algebras, Trans. American Math. Soc., 102 (1962), 352-372.

10. Raeburn, I. and Taylor, J.L., Hochschild cohomology and perturbations of Banach algebras, Preprint.

TWO SET-THEORETIC PROBLEMS IN TOPOLOGY

I. Juhász
Mathematical Institute of the
Hungarian Academy of Sciences
1053.Budapest, Hungary

The aim of this lecture is to draw your attention to some particular problems, rather than giving a survey of an area. I believe that the eventual solutions of these problems will significantly contribute to the progress of set-theoretic topology.

§.1. ON THE NUMBER OF OPEN SETS

For any topological space X I denote by $o(X)$ the number of all open subsets of X, i.e. the cardinality of the topology of X. J. de Groot raised the following problem in [3] : If X is an infinite Hausdorff space is $o(X)$ necessarily of the form $2^{\varkappa}$? He has observed that this is so for metric spaces. In [5] it has been shown that the answer to de Groot's question is affirmative if GCH and the non-existence of inaccessible cardinals are assumed.

On the other hand it follows from results in [6],[7] and [8] that in some models of set theory there are very good topological spaces such that e.g. $2^{\omega} < o(X) < 2^{\omega_1}$. In fact these spaces can be chosen as hereditarily separable topological groups or as regular and hereditarily Lindelöf. These results leave open the following problem:

1.1. PROBLEM. Let X be an infinite T_2 space. Is $o(X)^{\omega}=o(X)$?

The naturalness of this question is accentuated by a known result of R.S. Pierce and B. Efimov (cf. [2] and [15]) saying that the cardinality $\varkappa$ of an infinite complete Boolean algebra (i.e. the number of all regular open sets in a space) always satisfies $\varkappa^{\omega}=\varkappa$.

In what follows we present several partial results concerning problem 1.1. We shall be frequently using the following results on cardinal exponentiation (see e.g. [11]):

1.2. PROPOSITION. Let $\varkappa$ and λ be infinite cardinals. Then

A) $(\varkappa^{+})^{\lambda}=\varkappa^{+}\cdot\varkappa^{\lambda}$;

B) $\lambda < cf(\varkappa)$ implies $\varkappa^{\lambda}=\sum\{\alpha^{\lambda}:\alpha < \varkappa\}$;

C) if $\lambda=cf(\varkappa)$ and $\varkappa=\sum\{\varkappa_\nu:\nu\in\lambda\}$, where $\varkappa_\nu < \varkappa$ for $\nu\in\lambda$, then

$$\varkappa^\lambda = \Pi\{\varkappa_\nu : \nu\in\lambda\};$$

D) if $cf(\varkappa) < \lambda < \varkappa$, then

$$\varkappa^\lambda = (\sum\{\alpha^\lambda : \alpha < \varkappa\})^{cf(\varkappa)}$$

The next result also concerns cardinal exponentiation, and as far as I know it is new. It will play a crucial role in the proof given below.

Let $\varkappa$ and λ be cardinals, the power $\varkappa^\lambda$ is called a *jump*, if $\varkappa,\lambda \geq \omega$, $\alpha < \varkappa$ implies $\alpha^\lambda < \varkappa^\lambda$ and $\beta < \lambda$ implies $\varkappa^\beta < \varkappa^\lambda$.

1.3. LEMMA. If $\varkappa^\lambda$ is a jump, then $\lambda=cf(\varkappa)$.

Proof. First we show that $\alpha < \varkappa$ implies $\alpha^\lambda < \varkappa$. Indeed $\alpha^\lambda \geq \varkappa$ would imply $(\alpha^\lambda)^\lambda=\alpha^\lambda\geq\varkappa^\lambda$, a contradiction. In particular we obtain $2^\lambda < \varkappa$, hence $\lambda < \varkappa$.

Now assume that $\lambda < cf(\varkappa)$. Then by 1.2 *B)* and the above we have

$$\varkappa^\lambda = \sum\{\alpha^\lambda : \alpha < \varkappa\} \leq \varkappa=\varkappa^1 < \varkappa^\lambda$$

again a contradiction.

Finally, $cf(\varkappa) < \lambda$ would imply by 1.2 *D)* that

$$\varkappa^\lambda = (\sum\{\alpha^\lambda : \alpha < \varkappa\}^{cf(\varkappa)} \leq \varkappa^{cf(\varkappa)} < \varkappa^\lambda,$$

which is impossible. Thus, indeed, we must have $\lambda=cf(\varkappa)$.

The following three simple statements will be often used without mention in what follows. Their proofs are left to the reader.

1.4. LEMMA. Let R be an arbitrary topological space.

1) If $R=\cup\{R_i : i\in I\}$, then $o(R) \leq \Pi\{o(R_i) : i\in I\}$.

2) If $\{R_i : i\in I\}$ is a disjoint family of non-empty open subspaces of R, then

$$o(R) \geq \Pi\{o(R_i) : i\in I\}.$$

3) If there is a discrete subspace of cardinality $\varkappa$ in R, then $o(R) \geq 2^\varkappa$.

Let us denote by $\mathcal{H}$ the class of all strongly Hausdorff spaces (cf. [13]). Our next result, as we shall see, makes it very probable that the answer to problem 1.1 is affirmative at least for spaces in $\mathcal{H}$.

1.5. THEOREM. Let $\varkappa$ be a cardinal such that $o(X)=\varkappa$ for some infinite $X\in\mathcal{H}$ and $\varkappa < \varkappa^{\omega}$. Then there is a cardinal β with the following properties *(i)* - *(iv)*:

(i) $\omega < cf(\beta) = \gamma < \beta$;

(ii) $(\forall\alpha < \beta)(\alpha^{\gamma} < \beta)$;

(iii) $\beta^{\gamma} > \beta^{(\omega)}$ (=the ωth successor of β);

(iv) $\varkappa \geq \beta^{(\omega)}$.

Proof. Let λ be the smallest cardinal such that $\lambda^{\omega} > \varkappa$. Since $\lambda \leq \varkappa$, the power λ^{ω} is clearly a jump, hence by 1.3 we have $\omega=cf(\lambda)$. Moreover $\varkappa=o(X) > 2^{\omega}$ implies $\lambda > \omega$.

For any $p\in X$ let us put

$$\sigma(p,X) = \min\{o(U) : p\in U, \quad U \text{ open in } X\}$$

and

$$\sigma = \sigma(X) = \sup\{\sigma(p,X) : p\in X\}.$$

Since $X\in\mathcal{H}$ there can only be finitely many points $p\in X$ such that $\sigma(p,X) \geq \lambda$, for otherwise X would contain a disjoint family $\{U_n : n\in\omega\}$ of open sets with $o(U_n) \geq \lambda$ for all $n\in\omega$, and thus

$$o(X) \geq \lambda^{\omega} > \varkappa = o(X)$$

would follow. On the other hand, throwing away finitely many points from X will clearly not change $o(X)$, hence we can assume that $\sigma(p,X) < \lambda$ for each $p\in X$.

Now we claim that in fact $\sigma < \lambda$ must be valid. Assume, on the contrary, that $\sigma=\lambda$. Since λ can be written as $\lambda=\sum\{\lambda_n : n\in\omega\}$, where $\lambda_n < \lambda$ for $n\in\omega$, then we can pick for $n\in\omega$ distinct points $p_n\in X$ such that $\sigma(p_n,X) > \lambda_n$, moreover using $X\in\mathcal{H}$ we can also assume that each p_n has a neighbourhood U_n so that the family $\{U_n : n\in\omega\}$ is disjoint. However this implies, by 1.2 *c)*,

$$o(X) \geq \Pi\{o(U_n) : n\in\omega\} \geq \prod_{n\in\omega} \lambda_n = \lambda^{\omega} > \varkappa,$$

a contradiction.

Next we show that $|X| \leq \sigma^+$. Indeed, every $p \in X$ has an open neighbourhood $U(p)$ such that $|U(p)| \leq o(U(p)) \leq \sigma$. Hence if $|X| > \sigma^+$ were true then $U(p)$ would be a set-mapping which satisfies the conditions of Hajnal's theorem (cf [4] or [13]), hence a free set $D \subset X$ with $|D| = |X|$ would exist for $U(p)$. However this subspace D is clearly discrete, consequently $\varkappa = o(X) = 2^{|X|}$, which is of course impossible.

Now consider the above defined open cover $\mathcal{U} = \{U(p) : p \in X\}$ of X, then $|\mathcal{U}| \leq \sigma^+$. Let τ denote the smallest cardinal for which X *does not* contain a discrete subspace of cardinality τ. As is shown in [9], then every closed subset $F \subset X$ can be obtained in the following form

$$F = (F \cap (\cup \mathcal{U}_F)) \cup \bar{S}_F ,$$

where $\mathcal{U}_F \in [\mathcal{U}]^{<\tau}$ and $S_F \in [X]^{<\tau}$. An easy calculation shows then that

$$\lambda \leq o(X) \leq (\sigma^+)^{\underset{\smile}{\tau}}.$$

Since $X \in \mathcal{H}$, then 3.3 of [13] implies $cf(\tau) > \omega$. From this and $cf(\lambda) = \omega$ it follows then that there is a cardinal $\rho < \tau$ with $(\sigma^+)^\rho > \lambda$. Let γ be the smallest cardinal with $(\sigma^+)^\gamma > \lambda$ and then β be smallest such that $\beta^\gamma > \lambda$. Then $\beta \leq \sigma^+ < \lambda$, hence $\gamma > \omega$ by the choice of λ. Moreover $\gamma < \tau$, hence X contains a discrete subspace of size γ, consequently $o(X) \geq 2^\gamma = \gamma^\gamma$ and thus $\beta^\gamma \geq \lambda^\omega > o(X)$ implies $\beta > \gamma$. In particular β and γ are infinite, hence the power β^γ is a jump and therefore $\gamma = cf(\beta)$. Now it is obvious that β satisfies conditions *(i) - (iv)*.

As an immediate corollary we obtain the main result of [9] saying that if $X \in \mathcal{H}$ and $o(X) < \omega_{\omega_1+\omega}$ then $o(X)^\omega = o(X)$. Indeed, this is obvious since ω_{ω_1} is the smallest cardinal which satisfies *(i)*. However our result says much more than this. Indeed, the consistency of the existence of a cardinal satisfying *(i) - (iii)* has only been established by M. Magidor [14] with the help of some enormously large (so called strongly compact) cardinals. Moreover by some very recent result of R. Jensen [12] the existence of such a β implies that measurable cardinals exist in some inner models of set theory. This shows that constructing a "counterexample" to 1.1 would require some very sophisticated new method in axiomatic set theory.

It is natural to ask now wether an affirmative answer to 1.1 could obtained for special classes of Hausdorff spaces. Our next two results are of this form. Let P denote the class of all hereditarily paracompact T_3 spaces.

1.6. THEOREM. If $X \in P$ and $|X| \geq \omega$, then $o(X)=o(X)^{\omega}$.

Proof. Suppose, on the contrary, that $\varkappa=o(X) < \varkappa^{\omega}$. Similarly as in the proof of 1.5 we let λ be the smallest cardinal whose ω^{th} power exceeds $\varkappa$. Then $cf(\lambda)=\omega < \lambda \leq \varkappa$. We can of course assume that for all $Y \subset X$ with $o(Y)=\varkappa$ we have $\sigma(Y)=\sigma(X)$. Since $P \subset \mathcal{H}$, and the class P is hereditary, the same argument as in the proof of 1.5 yields then that $\sigma=\sigma(X) < \lambda$. Put $\rho=\min\{\alpha : \sigma^{\alpha} > \lambda\}$. By the choice of λ then $\rho > \omega$. The following claim is the crux of the proof.

Claim. Let $\langle \varkappa_{\xi} : \xi < \rho \rangle$ be a sequence of cardinals such that $\varkappa_{\xi} < \sigma$ for every $\xi < \rho$. Then there is a disjoint family $\{G_{\xi} : \xi < \rho\}$ of sets open in X such that $\varkappa_{\xi} < o(G_{\xi})$ for each $\xi < \rho$. In particular $\Pi\{\varkappa_{\xi} : \xi < \rho\} \leq o(X)=\varkappa$.

Proof of the claim. Clearly we have a locally finite open cover U of X for which $o(\bar{U}) \leq \sigma$ for every $U \in$ U. Now we define by transfinite induction for $\xi < \rho$ open sets $G_{\xi} \subset X$ and $U_{\xi} \in$ U such that $G_{\xi} \subset U_{\xi}$. Suppose that $\eta < \rho$ and G_{ξ}, U_{ξ} have been defined for $\xi < \eta$. Then

$$o(\cup\{\bar{U}_{\xi} : \xi < \eta\}) \leq \sigma^{|\eta|} < \lambda \leq \varkappa,$$

hence for $Y=X \setminus \cup\{\bar{U}_{\xi} : \xi < \eta\}$ we have $o(Y)=\varkappa$. Since U is locally finite Y is open, moreover $\sigma(Y)=\sigma$ by our assumption. Thus there is $p \in Y$ for which $\sigma(p,Y)=\sigma(p,X) > \varkappa_{\eta}$. Now pick $U_{\eta} \in$ U such that $p \in U_{\eta}$, and put $G_{\eta}=Y \cap U_{\eta}$. Then $p \in G_{\eta}$ implies $o(G_{\eta}) \geq \sigma(p,X) > \varkappa_{\eta}$, and clearly $\xi < \eta$ implies $G_{\xi} \cap G_{\eta}=\emptyset$. The claim is thus proven.

An immediate consequence of this claim is that $\tau < \sigma$ implies $\tau^{\rho} \leq \varkappa$, and thus $\tau^{\rho} < \lambda$ as well (indeed, $\tau^{\rho} \geq \lambda$ would imply $\tau^{\rho} \geq \lambda^{\omega} > \varkappa$). Consequently the power σ^{ρ} is a jump, hence $\rho=cf(\sigma)$ by 1.3. Now write $\sigma=\sum\{\varkappa_{\xi} : \xi < \rho\}$, where $\varkappa_{\xi} < \sigma$ for each $\xi < \rho$. Applying the claim to the sequence $\langle \varkappa_{\xi} : \xi < \rho \rangle$ we get a disjoint open family $\{G_{\xi} : \xi < \rho\}$ such that $o(G_{\xi}) > \varkappa_{\xi}$ for $\xi < \rho$. But then by 1.2 c)

$$\sigma^{\rho}=\Pi\{\varkappa_{\xi} : \xi < \rho\} \leq \Pi\{o(G_{\xi}) : \xi < \rho\} \leq o(X) = \varkappa < \lambda^{\omega},$$

while clearly $\sigma^{\rho} > \lambda$ implies $\sigma^{\rho} \geq \lambda^{\omega}$, a contradiction, which completes our proof.

Now let G be the class of all T_2 topological groups.

1.7. THEOREM. Let $G \in$ G, $|G| \geq \omega$. Then $o(G) = o(G)^{\omega}$.

Proof. Let e denote the unit element of G, V be the neighbourhood filter of e in G, and put $\sigma = \sigma(e,G) = \sigma(G)$. We have to distinguish two cases:

Case a. There is $V \in$ V such that $o(V) = \sigma$ and finitely many left tranlates of V cover G, i.e. there is a finite set $A \subset G$ for which $G = \cup\{aV: a \in A\}$. Clearly then $o(G) \leq \Pi\{o(a \cdot V): a \in A\} = \sigma$, while G contains an infinite disjoint family $\{H_n: n \in \omega\}$ of non-empty open sets, hence by $o(H_n) \geq \sigma$ we have $o(G) \geq \sigma^{\omega}$ and consequently $o(G) = o(G)^{\omega}$.

Case b. There is no $V \in$ V as in case a. Let $U \in$ V be arbitrary with $o(U) = \sigma$ and pick a symmetric neighbourhood $V \in$ V such that $V^2 \subset U$. Consider $A \subset G$ such that $\{aV: a \in A\}$ forms a maximal disjoint family of left translates of V. We claim that $\cup\{aU: a \in A\} = G$. Indeed for any $x \in G$ there is $a \in A$ with $(xV) \cap (aV) \neq \emptyset$, hence there are $v_1, v_2 \in V$ such that $xv_1 = av_2$. Then $x = av_2 v_1^{-1}$, and $v_2 v_1^{-1} \in U$ implies $x \in aU$.

Thus by our assumption $|A| = \alpha \geq \omega$, and obviously

$$o(G) \leq \Pi\{o(aU): a \in A\} = \sigma^{\alpha}$$

on the one hand and

$$o(G) \geq \Pi\{o(aV): a \in A\} = \sigma^{\alpha}$$

on the other. But then $o(G) = \sigma^{\alpha} = (\sigma^{\alpha})^{\omega}$.

I would like to mention at the end of this section the following problem.

1.8. PROBLEM. If X is an infinite compact T_2 space, is $o(X) = o(X)^{\omega}$?

§.2. OMITTING CARDINALS BY COMPACT SPACES

The problems considered in this section are motivated by [10], where the following question is investigated *(under GCH)*: does every Lindelöf space of cardinality ω_2 contain a Lindelöf subspace of cardinality ω_1? There it is also shown that any uncountable compact T_2 space contains a Lindelöf subspace of cardinality ω_1. This leads naturally to the following definition.

2.1. DEFINITION. The compact T_2 space X is said to omit the infinite cardinal $\varkappa$ if $|X| > \varkappa$ and X contains no closed *(=compact)* subspace of cardinality $\varkappa$.

EXAMPLE 1. It is well-known that βN omits every infinite $\varkappa < 2^{2^{\omega}}$.

The following example is due to E.van Dowen and it is included here with his kind permission.

EXAMPLE 2. Let λ be a strong limit cardinal with $cf(\lambda)=\omega$, and X be a compact T_2 space such that $|X| > \lambda$ and every countable discrete subset of X is C^*-embedded (e.g. X is an F-space, cf [1]). Then X omits λ.

It is enough to show that $Y \subset X$ and $|Y|=\lambda$ implies $|\bar{Y}| \geq \lambda^{\omega}(>\lambda)$. In fact we can restrict ourselves to discrete subspaces, because by 3.2 of [13] any such Y contains a discrete subspace of size λ. Now let A be a family of almost disjoint ω-element subsets of Y with $|\mathsf{A}|=\lambda^{\omega}$ *(cf.* [1] or [16]). It is easy to see that if $A,B \in \mathsf{A}$ and $A \neq B$ then no limit point of A is a limit point of B, hence clearly $|\bar{Y}| \geq |\mathsf{A}|=\lambda^{\omega}$.

The above two examples tell all what is known about cardinals that *can be* omitted by a compact T_2 space.

In what follows we always assume GCH. Next we want to formulate a result showing that the omitting of cardinals by compact spaces is subject to some strict limitations. First however we prove a lemma which is interesting in itself.

2.2. LEMMA. *(GCH)* Suppose X is compact T_2 and omits $\varkappa^+$; then there is a closed subspace $F \subset X$ with a point $p \in F$ such that $\chi(p,F)=\varkappa^+$.

Proof. We can of course assume that X has a dense subset of size $\varkappa^+$ and that $\chi(p,X) \neq \varkappa^+$ for each $p \in X$. Then by 2.20 of [13] we have $|\{p \in X: \chi(p,X) \leq \varkappa\}| \leq (\varkappa^+)^{\varkappa}=\varkappa^+$, hence $\{p \in X : \chi(p,X) \geq \varkappa^{++}\}$ is a $G_{\varkappa^+}$-set in X *(i.e. the intersection of* $\varkappa^+$ *open sets)*, and therefore it contains a closed non-empty $G_{\varkappa^+}$-set Z. It is easy to see that

$\chi(p,Z) \geq \varkappa^{++}$ is valid for all $p\in Z$, hence we can apply proposition 2 of [10] to find a set $A\subset Z$ with $|A|=2^{\varkappa}=\varkappa$ such that $|\bar{A}| > \varkappa$. Since X omits $\varkappa^{+}$, we actually have $|\bar{A}| > \varkappa^{+}$ as well. Now let $F=\bar{A}$. Then $w(F) \leq 2^{|A|}=\varkappa^{+}$, hence $\chi(p,F) \leq \varkappa^{+}$ for every $p\in F$. On the other hand by 2.20 of [13] again $|\{p\in F: \chi(p,F) \leq \varkappa\}| \leq \varkappa^{\varkappa}=\varkappa^{+} < |F|$, hence we must have a point $p\in F$ with $\chi(p,F)=\varkappa^{+}$.

2.3. THEOREM. *(GCH)* A compact T_2 space X cannot omit both $\varkappa^{+}$ and $\varkappa^{++}$.

Proof. Suppose that $|X| > \varkappa^{++}$ and X omits $\varkappa^{+}$. Then by lemma 2.2 there is a closed subsapce $F\subset X$ with a point $p\in F$ such that $\chi(p,F)=\varkappa^{+}$. One can easily construct then a strictly decreasing sequence $\{K_\nu: \nu < \varkappa^{+}\}$ of closed subsets of F such that $\cap\{K_\nu: \nu<\varkappa^{+}\}=\{p\}$. For each $\nu < \varkappa^{+}$ pick a point $p_\nu\in K_\nu\setminus K_{\nu+1}$. Clearly then

$$\overline{\{p_\nu: \nu < \varkappa^{+}\}} = \bigcup_{\alpha<\varkappa^{+}} \overline{\{p_\nu: \nu < \alpha\}} \cup \{p\}.$$

If there is an $\alpha < \varkappa^{+}$ such that $\overline{\{p_\nu: \nu < \alpha\}}$ has cardinality $\varkappa^{++}$, we are done. If not, i.e. if $|\overline{\{p_\nu: \nu < \alpha\}}| \leq \varkappa^{+}$ for each $\alpha < \varkappa^{+}$, then clearly $|\overline{\{p_\nu: \nu < \varkappa^{+}\}}|=\varkappa^{+}$, which is impossible since X omits $\varkappa^{+}$.

Finally we mention the following simple result.

2.4. THEOREM. *(GCH)* A compact T_2 space cannot omit both ω and ω_2.

Proof. This follows immediately from theorem 2 of [10] which says that if a compact T_2 space X omits ω then $|X| \geq 2^{\omega_1}$.

The following two simplest questions remain open:

2.5. PROBLEM *(GCH)* Can a compact T_2 space omit ω_2, or ω_1 and ω_3?

REFERENCES

[1] W.W. Comfort and S. Negrepontis, The theory of ultrafilters, Springer-Verlag, New York-Heidelberg-Berlin, 1974.

[2] B. Efimov, On extremally disconnected bicompacta, Dokl. Akad. Nauk SSR, 172*(1967)*, 771-774. *(in Russian)*

[3] J. de Groot, Discrete subspaces of Hausdorff spaces, Bull. Acad. Pol. Sci.*(13 1965)*, 537-544.

[4] A. Hajnal, Proof of a conjecture of S. Ruziewicz, Fund. Math. 50*(1961/62)*, 123-128.

[5] A. Hajnal and I. Juhász, Discrete subspaces of topological spaces, II, Indog. Math. 31*(1969)*, 18-30.

[6] A.Hajnal and I. Juhász, A consistency result concerning hereditarily α-Lindelöf spaces, Acta Math. Acad. Sci. Hung.

[7] A.Hajnal and I. Juhász, A consistency result concerning hereditarily α-separable spaces, Indag. Math. 35(1973), 301-307

[8] A. Hajnal and I. Juhász, A separable normal topological group need not be Lindelöf, Gen. Top. Appl. 6(1976), 199-205

[9] A. Hajnal and I. Juhász, On the number of open sets, Ann. Univ. Sci. Budapest, 16(1973), 99-102

[10] A. Hajnal and I. Juhász, Remarks on the cardinality of compact spaces and their Lindelöf subspaces, Proc. AMS (to appear)

[11] T. Jech, Lectures in set theory, Springer Lecture Notes in Math., Vol. 217, Berlin-Heidelberg-New York, 1971

[12] R.B. Jensen and K. Devlin, Marginalia to a Theorem of Silver. (to appear)

[13] I. Juhász, Cardinal functions in topology, Math. Centre Tract. No. 34, Amsterdam, 1971

[14] M. Magidor, On the singular cardinals problem, I. (to appear)

[15] R.S. Pierce, On a set-theoretic problem, Proc. Amer. Math. Soc., 9(1958), 892-896

[16] A. Tarski, Sur la décomposition des ensembles en sous-ensembles presque dijoints, Fund. Math. 12(1928), 188-205.

CATEGORY, BOOLEAN ALGEBRAS AND MEASURE

D. Maharam
University of Rochester
Rochester, N. Y., U. S. A.

Introduction

It should be said at once that the "category" in the title refers to Baire category. A topological measure space X will have three naturally-arising complete Boolean algebras: the algebras $\mathcal{A}_r(X)$ of regular open sets, $\mathcal{A}_c(X)$ of Borel sets modulo first category sets, and $\mathcal{A}_m(X)$ of measurable sets modulo null sets. While $\mathcal{A}_m(X)$ is obviously the algebra of greatest interest to analysts, $\mathcal{A}_c(X)$ (the "category algebra" of X in the terminology of Oxtoby [10]) is also of considerable interest to them. It turns out, however, that $\mathcal{A}_c(X)$ does not behave very well (under product formation, for instance) unless X is "nice", in which case $\mathcal{A}_c(X)$ is the same as $\mathcal{A}_r(X)$. Thus it pays to prove general theorems about the better-behaved $\mathcal{A}_r(X)$ rather than $\mathcal{A}_c(X)$, even though one may be more interested in the latter.

Accordingly we begin by discussing $\mathcal{A}_r$, in § 1. In § 2 and § 3 we consider $\mathcal{A}_c$ and $\mathcal{A}_m$ respectively. In §§ 4-6 we compare and contrast the behavior of $\mathcal{A}_c$ and $\mathcal{A}_m$ with respect to problems concerning liftings, completions and mappings from representation spaces. Finally in § 7 we apply some of the results of previous sections to construct a "completion" for C(X). The unifying thread connecting these topics is simply that they have arisen in the course of the author's recent work. Much of what follows may well be known, but (apart from the references given below) I have not found most of it in the literature.

I am grateful to A. H. Stone for some helpful discussions.

1. The regular open algebra

For an arbitrary topological space X we write $\mathcal{A}_r(X)$ for the family of regular open subsets of X; this (ordered by set inclusion) is well known to be a complete Boolean algebra, the supremum of a family of regular open sets being the interior of the closure of their union. (The infimum of a finite number of regular open sets is their intersection.) We note that every complete Boolean algebra $\mathcal{A}$ arises as the regular open algebra of some (compact, Hausdorff, extremally disconnected) space X - namely, the Stone representation space of $\mathcal{A}$, which we shall denote by $R(\mathcal{A})$. (See [13, p. 117].)

We say that two spaces X, Y are "regular open equivalent", and write $X \sim_r Y$, to mean that $\mathcal{A}_r(X)$ and $\mathcal{A}_r(Y)$ are isomorphic. Thus, for example, $X \sim_r R(\mathcal{A}_r(X))$ for all X. Oxtoby [10] has given a method for constructing all spaces Y for which $X \sim_r Y$; but it is (inevitably) not easily applied in particular cases - for, from the above remark, an effective method here would imply the classification of all complete Boolean algebras. However, Oxtoby obtains a striking consequence (though it is easily proved directly):

(1) If D is a dense subset of X, then $D \sim_r X$.

As this shows, r-equivalent spaces can be very different topologically. Thus, for example, if X is a Tychonoff space, then $X \sim_r \beta X$. Again, let K denote $\{0\} \cup \{n^{-1}: n \in N\}$, where N is the set of positive integers; then $K \sim_r \beta N$ (because both have dense subsets homeomorphic to N), despite the disparity in their cardinals, and despite the fact that both are compact.

However, r-equivalence does preserve some topological properties. The following instances were obtained jointly by A.H. Stone and myself. First, the "density character" $\delta(X)$ is defined as usual as the smallest cardinal of a dense subset of X. Define the "dense density character" $\delta\delta(X)$ to be the smallest cardinal d such that every dense subset of X has a dense subset of cardinal $\leq d$. (Clearly $\delta(X) \leq \delta\delta(X)$; there need not be equality.) Then:

(2) If $X \sim_r Y$ then $\delta(X) = \delta(Y)$.

(3) If $X \sim_r Y$ and X, Y are compact Hausdorff, then $\delta\delta(X) = \delta\delta(Y)$.

(4) If X and Y are non-empty separable metric spaces (or, more generally, are regular T_1 first countable spaces) without isolated points, then $X \sim_r Y$.

For X and Y, in (4), will have dense subspaces homeomorphic to the space Q of rationals, as follows from [12]. Note that (4) applies, for example, to the Sorgenfrey line and plane.

To formulate a more inclusive result that allows for isolated points, write $\mathcal{J}(X)$ = the set of all isolated points of X, $\mathcal{D}(X) = X - \mathrm{Cl}(\mathcal{J}(X))$. Note that the isolated points of a regular T_1 space X are precisely the atoms of $\mathcal{A}_r(X)$, and that $\mathcal{D}(X)$ is precisely the complement, in $\mathcal{A}_r(X)$, of their supremum. Hence a regular open equivalence between X and Y induces a one-one correspondence between $\mathcal{I}(X)$ and $\mathcal{I}(Y)$, and also a regular open equivalence between $\mathcal{D}(X)$ and $\mathcal{D}(Y)$. Conversely, we have (taking the metrizable case for simplicity of statement, and writing $|E|$ for the cardinal of E):

(5) If X and Y are separable metric spaces, and if $|\mathcal{I}(X)| = |\mathcal{I}(Y)|$ and $\mathcal{D}(X)$, $\mathcal{D}(Y)$ are either both empty or both non-empty, then $X \sim_r Y$.

For, as in (4), $\mathcal{D}(X)$ and $\mathcal{D}(Y)$ will have homeomorphic dense subspaces, and these, together with $\mathcal{I}(X)$ and $\mathcal{I}(Y)$, provide regular open equivalent dense subspaces of X and Y, to which we apply (1).

A fairly straightforward argument will also prove:

(6) If $X_\lambda \sim_r Y_\lambda$ (for all $\lambda \in \Lambda$) then $\prod_\lambda X_\lambda \sim_r \prod_\lambda Y_\lambda$.

From this and the foregoing we see that, for example, if k is any uncountable cardinal, the spaces 2^k, N^k, R^k, I^k, $(\beta N)^k$ are all regular open equivalent.

Another way of looking at these results comes from the fact that in these cases (and some others) it is possible to give fairly simple characterizations of the Boolean algebras $\mathcal{A}_r(X)$. For instance, if X is as in (4) - we may as well say $X = I$, the unit interval - then $\mathcal{A}_r(X)$ is characterized, to within isomorphism, as being a complete non-atomic Boolean algebra with a countable σ-basis (see [1, p. 177]). From this a (more complicated) characterization of $\mathcal{A}_r(I^k)$ can be derived. Of course, if X has a dense discrete subset D, $\mathcal{A}_r(X)$ is isomorphic to the algebra $\mathcal{P}(D)$ of all subsets of D, for which characterizations are also known [14]. And in (5), $\mathcal{A}_r(X)$ is characterized (if $\mathcal{D}(X) \neq \emptyset$) as the direct sum of $\mathcal{A}_r(I)$ and $\mathcal{P}(\mathcal{I}(X))$.

Finally we mention the easily verified fact:

(7) If $f : X \to Y$ is a continuous open surjection, then f^{-1} gives an isomorphism of $\mathcal{A}_r(Y)$ onto a complete subalgebra of $\mathcal{A}_r(X)$.

2. The category algebra

Again let X be a topological space, $\mathcal{B}$ its family of Borel sets, $\mathcal{C}$ its family of sets of first category (in X). Let $\mathcal{B} + \mathcal{C}$ denote the family of sets differing from Borel sets by sets of first category. The "category algebra" $\mathcal{A}_c(X)$ is defined to be the quotient algebra $(\mathcal{B} + \mathcal{C})/\mathcal{C}$. As is well known, there is a natural homomorphism $f : \mathcal{A}_r(X) \to \mathcal{A}_c(X)$, which is an isomorphism if, and only if, X is a Baire space (that is, no non-empty open subset of X is of first category in itself - or, equivalently, in X). We define $X \sim_c Y$ to mean that $\mathcal{A}_c(X)$ and $\mathcal{A}_c(Y)$ are isomorphic. Thus, for locally compact Hausdorff spaces, and for complete metric spaces, $\mathcal{A}_r(X) = \mathcal{A}_c(X)$, and $\sim_c$ coincides with $\sim_r$.

In general, there is no implication between $\sim_c$ and $\sim_r$. For example, $Q \sim_r R$, by 1.(4), but $\mathcal{A}_c(Q) = \{0\} \neq \mathcal{A}_c(R)$. Again, $\{0\} \sim_c Q$, but $\mathcal{A}_r(\{0\}) \neq \mathcal{A}_r(Q)$. Nevertheless there is a sense in which the category algebra is reducible to the regular open algebra (and $\sim_c$ to $\sim_r$). For, given a space X , the union U^* of all its open sets of first category is, by a theorem of Banach [6, p. 82], also an open set of first category. Put $X^* = X - \overline{U^*}$; then X^* is a Baire space, and $\mathcal{A}_c(X)$ is isomorphic to $\mathcal{A}_c(X^*) = \mathcal{A}_r(X^*)$.

This shows that the assumption we shall usually make, when studying the category algebra, that the spaces involved are Baire, is not an enormous one. It enables us to transfer the results of the previous section to $\mathcal{A}_c$ and $\sim_c$; for instance, 1.(1) says that if D is a dense subset of X , and both D and X are Baire spaces (it suffices that D is Baire), then $D \sim_c X$. Of course, 1.(3) applies to $\sim_c$ as it stands. Note that the analogue of 1.(6) is complicated by the need to require that the product spaces too are Baire sets, which in general they need not be ([11], [16]).

Not every significant property of $\sim_c$ arises as a special case of one of $\sim_r$. A topological space X is said to be "residually Lindelöf" if every open cover $\mathcal{U}$ of X has a countable subsystem $U_1, U_2, \ldots$, such that $X - \bigcup_{n=1}^{\infty} U_n$ is of first category (see [5]). Say that X is "hereditarily residually Lindelöf" if every open subset of X is residually Lindelöf. Then we have:

(1) Suppose X and Y are Baire spaces and $X \sim_c Y$. Then if X is hereditarily residually Lindelöf, so is Y.

(More generally, an analogous definition can be given for "hereditarily residually $(\alpha - \beta)$ compact", and the analogous result will hold.) Thus, for instance, every open subset of $R(\mathcal{A}_c(I))$ will be residually Lindelöf. Note that the analogue of (1) for not necessarily Baire spaces such that $X \sim_r Y$ would be false - for instance when $X = R \times D$ and $Y = Q \times D$ with D an uncountable discrete space.

3. The measure algebra

Now assume that the topological space X also has a finite (or σ-finite) regular Borel measure μ; that is, μ is a non-negative, countably additive measure defined on the family $\mathcal{B}$ of Borel sets of X, with the property that $\mu(B) = \inf\{\mu(G) : G \text{ is open and } G \supset B\}$ for each $B \in \mathcal{B}$. Put $\mathcal{N} = \{E \subset X : \text{there exists } B \in \mathcal{B} \text{ such that } B \supset E$ and $\mu(B) = 0\}$. Then μ extends in the obvious way to the family $\mathcal{B} + \mathcal{N}$ of sets that differ from Borel sets by members of $\mathcal{N}$. We put $\mathcal{A}_m = (\mathcal{B} + \mathcal{N})/\mathcal{N}$. This too is a complete Boolean algebra, and it presents some analogies with $\mathcal{A}_c$. For instance, we can without much loss require that $\mu(G)$ be positive for every non-empty open set G, by replacing X by the complement of the union of all open sets of measure 0 (this union is of measure 0 because, since μ is σ-finite, $\mathcal{A}_m$ satisfies the countable chain condition). This would be the analogue of replacing X by the Baire space X^* in the previous section. Nevertheless there are some sharp differences between $\mathcal{A}_m$ on the one hand, and $\mathcal{A}_c$ and $\mathcal{A}_r$ on the other. Like $\mathcal{A}_r$, $\mathcal{A}_c$ can hardly be expected to have a simple explicit structure theory, for that would amount to a structure theory for all complete Boolean algebras. But $\mathcal{A}_m$ has a reasonably satisfactory structure theory (independent of the topological assumptions), as follows. Write $X \sim_m Y$ to mean that $\mathcal{A}_m(X)$ and $\mathcal{A}_m(Y)$ are isomorphic. Then [7] given X we have $X \sim_m Y$ where Y is the discrete union of countably many measure spaces, each of which is either an atom or (to within a constant scaling factor) a product I^k of copies of the unit interval I, with product Lebesgue measure.

Another difference is that $\mathcal{A}_m$ has the property (a consequence of the regularity of μ and of Urysohn's Lemma):

(1) If X is normal (qua topological space), each measure class contains a Baire set.

(Here, as usual, the Baire sets are the σ-field generated by the zero-sets.) The analogue of (1) for $\mathcal{A}_c$ is false, in general, even for compact Hausdorff spaces, as is shown by the following example (pointed out to me by A. H. Stone). Take X to be the usual space of ordinals $\leq \omega_1$, and split the non-limit ordinals into two complementary cofinal sets, say E and F. Both E and F are open, hence Borel; but neither can differ from a Baire set by a first category set.

Nevertheless, in every product of separable metric spaces (I^k, for instance) it can be shown that every regular open set <u>is</u> a Baire set; thus in this case each category class (of a Borel set modulo first category) does contain a Baire set. It would be interesting to know (a) for what spaces every Borel set differs from a Baire set by a set of first category, (b) for what spaces all regular open sets are Baire (or, more specifically, are co-zero).

We observe that, in $\mathcal{A}_c(X)$, each category class a contains a largest open set $G(a)$ (namely, the union of all open sets in the class) and a smallest closed set $F(a)$. If X is a Baire space, $G(a)$ is the unique regular open set in a, and $F(a)$ is the unique regular closed set in a, and we have $F(a) = Cl(G(a)), G(a) = Int(F(a))$. The analogue for $\mathcal{A}_m(X)$ fails; in general, a measure class a will contain neither an open set nor a closed set. However, <u>if</u> the measure class a contains an open set, it contains a largest one, say $G_1(a)$, and we call a an "open class". Similarly a "closed class" a is one that contains a closed set, and hence a smallest closed set, say $F_1(a)$. The "<u>ambiguous classes</u>" are defined to be those that are both open and closed. (This notion has been considered independently by S. Graf, in unpublished work.) If we assume (without essential loss, as remarked above) that each non-empty open set in X has positive measure, then we have $F_1(a) = Cl(G_1(a))$ and $G_1(a) = Int(F_1(a))$ for all ambiguous classes a, in analogy with the situation in $\mathcal{A}_c(X)$. We shall make use of the ambiguous measure classes in § 6 below.

4. <u>Liftings</u>

Suppose $\mathcal{E}$ is an arbitrary Boolean algebra, and $\mathcal{J}$ is an arbitrary ideal in $\mathcal{E}$. Let $\mathcal{A}$ be the factor algebra $\mathcal{E}/\mathcal{J}$. A "lifting" of $\mathcal{A}$ is a homomorphism h of $\mathcal{A}$ into $\mathcal{E}$ (qua finitely additive Boolean algebras; h need not preserve infinite operations, even if they are available), such that $h(A) \in a$ for all $a \in \mathcal{A}$. Suppose in particular that $\mathcal{E}$ is an algebra of subsets of a space X; then a "strong lifting" is one with the property that whenever $G \in \mathcal{E}$ is an open set in X with g (say) as its class mod $\mathcal{J}$, then $h(g)$ is an open set containing G.

The following theorem seems to be generally known, though I have not seen it in print in exactly this form. It follows easily from a theorem of Graf [4]; and independent, unpublished proofs have been obtained by J. P. R. Christensen and by myself.

(1) If X is a Baire space, the category algebra $\mathcal{A}_c(X) = (\mathcal{B} + \mathcal{C})/\mathcal{C}$ always has a strong lifting.

The proof of (1) is basically a Zorn's Lemma argument, taking the representative $h(a)$ to be intermediate between $G(a)$ and $F(a)$, in the notation of the previous section. It is (so far as I know) an open question whether one can always take $h(a)$ to be a Borel set.

Analogously, $\mathcal{A}_m = (\mathcal{B} + \mathcal{N})/\mathcal{N}$ always has a lifting [8]; the roles of $G(a)$ and $F(a)$ in the preceding are now taken by the sets of upper and lower density. Again, it is (so far as I know) an open question whether there is always a <u>strong</u> lifting for $\mathcal{A}_m$ (say if X is compact Hausdorff), assuming of course that the measure of every non-empty open set is positive. Perhaps the study of the "ambiguous classes" of § 3 may throw some light on this.

In the same order of ideas, we can ask under what conditions an automorphism h of $\mathcal{A} = \mathcal{E}/\mathcal{I}$, where $\mathcal{E}$ is an algebra of subsets of X, can be "realized" by a suitable point-transformation $f : X \to X$ (so that $h(a)$ is in the class of $f(E)$ for every E in the class a). We are concerned here with the cases $\mathcal{A} = \mathcal{A}_c(X)$ or $\mathcal{A}_m(X)$. Even for these, easy counterexamples show that X will have to be very special; "compact Hausdorff" is not enough. However, Choksi has shown [2] that when X is a compact Hausdorff group, then every automorphism of $\mathcal{A}_m(X)$ can be realized by a (both-ways measurable) bijection of X onto itself. On the other hand, not every automorphism of $\mathcal{A}_c(C)$, where C is the Cantor set, can be realized by a <u>homeomorphism</u>. The following provides an example. Choose a 2-sided limit point $\alpha \in C$, and put

$A = [0,\alpha] \cap C$, $B = [\alpha, 1] \cap C$, $U = [0, 1/3] \cap C$, $V = [2/3, 1] \cap C$.
Then $\mathcal{A}_c(U) = \mathcal{A}_r(U)$ and $\mathcal{A}_c(A) = \mathcal{A}_r(A)$ are isomorphic, and $\mathcal{A}_c(V)$, $\mathcal{A}_c(B)$ are isomorphic; and these isomorphisms combine to give an automorphism of $\mathcal{A}_c(C)$ that takes the class of U to the class of A. If this could be realized by a homeomorphism h, then $h(U)$ and A would be regular closed sets in the same category class, and would therefore coincide; but $h(U)$ is open, and A is not. It can be shown, however, that <u>every automorphism of $\mathcal{A}_c(C)$ can be realized by a homeomorphism of a dense G_δ subset of C onto itself</u>. This answers a question asked me by S. Kakutani, in conversation. I hope to publish the proof elsewhere.

5. Completions

Let $\mathcal{F}$ be an arbitrary Boolean algebra; consider its representation space $R(\mathcal{F})$, and put $\mathcal{F}^* = \mathcal{A}_c(R(\mathcal{F}))$ $(= \mathcal{A}_r(R(\mathcal{F})))$. Then $\mathcal{F}$, qua finitely additive algebra, is a subalgebra of the complete algebra $\mathcal{F}^*$ (that is, the natural embedding of $\mathcal{F}$ in $\mathcal{F}^*$ preserves finite infs and sups, but not in general infinite ones, even when they are available). Roughly speaking, $\mathcal{F}^*$ is the smallest complete algebra containing $\mathcal{F}$ in this sense; this is the content of the following theorem, which follows easily from one in [13, p. 141]:

(1) If Θ is an isomorphism (finitely additive) of $\mathcal{F}$ into a complete Boolean algebra $\mathcal{G}$, then there is a unique extension of Θ to an isomorphism Θ^* of $\mathcal{F}^*$ onto a (finitely additive) subalgebra of $\mathcal{G}$.

(Here $\Theta^*(\mathcal{F}^*)$, though itself necessarily a complete algebra, is guaranteed only to have its finite operations agree with those of $\mathcal{G}$.)

Now suppose μ is a finitely additive (non-negative, finite) measure on $\mathcal{F}$. Then μ extends to a countably additive measure on the family $\mathcal{B}$ of Borel sets of $R(\mathcal{F})$; and the corresponding measure algebra $\mathcal{G}$ is a complete Boolean algebra, to which (1) applies. This (with some elementary considerations) proves:

(2) μ has a unique extension to a finitely additive measure μ^* on $\mathcal{F}^*$; further, μ^* is reduced (that is, vanishes only for the zero element) if, nad only if, μ is.

It follows, for example, that

(3) there exists a finitely additive, finite reduced measure on $\mathcal{A}_c(I^k)$, where k is an arbitrary infinite cardinal.

For $\mathcal{A}_c(I^k) = \mathcal{A}_c(2^k)$ by 1.(6). Let $\mathcal{F}$ denote the finitely additive algebra formed by the open-closed sets in 2^k. The restriction of the usual Lebesgue product measure to $\mathcal{F}$ gives a suitable μ to which (2) applies. Here $R(\mathcal{F}) = 2^k$, and therefore $\mathcal{F}^* = \mathcal{A}_c(2^k)$.

Note that $\mathcal{A}_c(I^k)$ does not carry a countably additive reduced, finite measure [1, p. 186].

It would be good to have a structure theory for finitely additive measures similar to that (described in § 3) for countably additive ones; but this will not be easy. One conjecture might be that such a finitely additive measure algebra - say with a reduced, nonatomic, finite measure μ - might be isomorphic to a direct sum of terms of the form $\mathcal{A}_c(2^k)$, each with a suitable (finitely additive) measure. Unfortunately this is false, because it can be shown that this would imply that the "density measures" on $\mathcal{P}(N)$ (see [9]) would have liftings; and they don't.

6. Spaces as continuous images of representation spaces

Let X be a compact Hausdorff space, and denote by $\tilde{X}$ the representation space $R(\mathcal{A}_c(X))$ ($= R(\mathcal{A}_r(X))$). Gleason has observed [3] that the natural isomorphism between $\mathcal{A}_c(X)$ and $\mathcal{A}_c(\tilde{X})$ can be realized by a continuous surjection $\Theta : \tilde{X} \to X$. In fact, one can define, for each $\tilde{\alpha} \in \tilde{X}$ (so that $\tilde{\alpha}$ is an ultrafilter on $\mathcal{A}_c(X)$)

$$\{\Theta(\tilde{\alpha})\} = \bigcap \{F(a) : a \in \tilde{\alpha}\} ,$$

where (as in § 3) $F(a)$ is the smallest (regular) closed set in the class $a \in \mathcal{A}_c(X)$. Of course, it has to be checked (among other things) that this intersection really is a singleton.

An analogous theorem holds for the measure-algebraic case. Suppose μ is a measure on X, as in § 3 above, and suppose further that X is compact Hausdorff and that every non-empty open subset of X has positive μ-measure. The ambiguous measure classes (defined at the end of § 3) form a finitely additive subalgebra $\mathcal{F}$ of $\mathcal{A}_m$. Put $X^* = R(\mathcal{F})$; the measure μ may then be regarded as defined on the open-closed subsets of X^*. It can be extended, in a standard way, to a countably additive measure μ^* on the Borel sets of X^*.

<u>Theorem. There is a continuous surjection</u> $\Theta_0 : X^* \to X$ <u>that realizes an isomorphism between</u> (X^*, μ^*) <u>and</u> (X, μ).

In fact, one can define $\{\Theta_0(\alpha^*)\} = \bigcap \{F_1(a) : a \in \alpha^*\}$, where $F_1(a)$ is the smallest closed set in $a \in \mathcal{A}_m(X)$.

This theorem provides a relatively simple proof of a theorem of C. Ionescu Tulcea [15, p. 169]. Still assuming X compact Hausdorff, and that μ is positive for non-empty open sets, put $X' = R(\mathcal{A}_m(X))$. As before, the measure μ on X then gives a finitely additive measure on the open-closed subsets of X', and we extend this to a countably additive measure μ' on the Borel subsets of X'. The theorem in question asserts that (under the above hypotheses on X and μ) <u>there is a continuous measure-preserving surjection</u> $\Theta' : (X', \mu') \to (X, \mu)$. To see this, note that there is a natural continuous map $\phi : X' \to X^*$;

this follows from the fact that $X' = R(\mathcal{A}_m)$ and $X^* = R(\mathcal{F})$ where $\mathcal{F}$ is a (finitely additive) subalgebra of $\mathcal{A}_m$. Now take $\Theta' = \Theta_0 \circ \Phi$; it is not hard to verify that this works.

7. A completion for C(X)

We use $C(X)$, as usual, to denote the partially ordered linear space of all continuous real-valued functions on X. $C(X)$ is also, of course, a ring; but we are more concerned with its linear properties. Suppose that X is compact Hausdorff, and let Θ be the Gleason map from $\tilde{X} = R(\mathcal{A}_c(X))$ to X. Then Θ induces a linear-space isomorphism $\Theta^*: C(X) \to C(\tilde{X})$; and it is easy to see that Θ^* is also an order-isomorphism and a ring isomorphism. Now $C(\tilde{X})$, qua partially ordered set, is conditionally complete; that is, every bounded subset of $C(\tilde{X})$ has a least upper bound. (This follows from the fact that X is extremally disconnected.) Further, the image $\Theta^*(C(X))$ can be shown to be order-dense in $C(\tilde{X})$ (one first shows that if $f \in C(\tilde{X})$ is a characteristic function then there exists $g \in C(X)$ such that $0 \leq \Theta^*(g) \leq f$). Conversely, if $\psi : C(X) \to L$ is an arbitrary order-preserving linear-space isomorphism into a conditionally complete partially ordered linear space L , it can be shown that there is an order-preserving linear-space isomorphism $\Phi : C(\tilde{X}) \to L$ such that $\Phi \circ \Theta^* = \psi$. Thus, in a reasonable sense, $C(\tilde{X})$ _is the smallest conditionally complete partially ordered linear space containing_ $C(X)$.

If X and Y are compact Hausdorff spaces, and $X \sim_c Y$, then $\tilde{X} = \tilde{Y}$, so that $C(X)$, $C(Y)$ will have the same "completions", in the above sense. It would be interesting to know whether the converse is true.

Essentially the same construction can be applied to all completely regular T_1 spaces X (not necessarily compact). We replace $C(X)$ by the subring $C_{ub}(X)$ of all uniformly continuous bounded functions (uniformly continuous in the uniformity induced by the finite open covers of X). Then $C_{ub}(X) = C(\beta X)$, and we apply the previous considerations to βX. Since $\mathcal{A}_r(X) = \mathcal{A}_r(\beta X) = \mathcal{A}_c(\beta X)$, the completion of $C_{ub}(X)$ will still be $C(\tilde{X})$ where now $\tilde{X} = R(\mathcal{A}_r(X))$.

Returning to the compact case, we note that the function space $C(\tilde{X})$ can be described more directly in terms of suitable classes of functions on X. Let $D(X)$ denote the set of all (real-valued) functions that are continuous and bounded (and defined) on residual subsets of X. Identify two functions in $D(X)$ if they agree on a residual set. This produces a partially ordered linear space $\tilde{D}(X)$.

<u>Theorem. If</u> X <u>is compact Hausdorff, then</u> $C(\tilde{X}) = \tilde{D}(X)$, <u>to within a natural isomorphism</u>.

The isomorphism here is such that, for each $f \in C(X)$, the class $\tilde{f}$ (of f mod first category) in $\tilde{D}(X)$ corresponds to $g \in C(\tilde{X})$ where $g = f \circ \Theta$, Θ being the Gleason map. The proof depends on the fact that, because of the extremal disconnectedness of $R(\mathcal{A}_c(X))$, each real-valued function on X that is continuous when restricted to a residual set, is equal (mod first category) to one that is continuous on all of X.

It can be shown that, if X is compact and Hausdorff and satisfies the countable chain condition (i. e., has no uncountable family of pairwise disjoint open sets), then $\tilde{D}(X)$ is identical with the set of all bounded "analytically representable" functions, modulo set of first category. (The analytically representable functions constitute the smallest family containing the continuous functions and closed under (pointwise) sequential limits.) The countable chain condition is not superfluous here, as is shown by essentially the same example as in § 3.

<u>References</u>

[1] G. Birkhoff: Lattice Theory. Amer. Math. Soc. Colloquium Pub. 25 (2nd ed.), New York 1948.

[2] J. Choksi: Measurable transformations on compact groups. Trans. Amer. Math. Soc. 184 (1973), 101-124.

[3] A. M. Gleason: Projective topological spaces. Illinois J. Math. 2 (1958), 482-489.

[4] S. Graf: Schnitte Boolescher Korrespondenzen und ihre Dualisierungen. Thesis, Universität Erlangen-Nürnberg 1973.

[5] J. H. B. Kemperman and D. Maharam: R^c is not almost Lindelöf. Proc. Amer. Math. Soc. 24 (1970), 772-773.

[6] K. Kuratowski: Topology vol. 1. Academic Press, New York 1966.

[7] D. Maharam: On homogeneous measure algebras. Proc. Nat. Acad. Sci. 28 (1942), 108-111.

[8] D. Maharam: On a theorem of von Neumann. Proc. Amer. Math. Soc. 9 (1958), 987-994.

[9] D. Maharam: Finitely additive measures on the integers. Sankhyā (to appear).

[10] J. C. Oxtoby: Spaces that admit a category measure. J. Reine Angew. Math. 205 (1961), 156-170.

[11] J. C. Oxtoby: Cartesian products of Baire spaces. Fund. Math. 49 (1961), 157-166.

[12] W. Sierpiński: Sur une propriété topologique des ensembles denses en soi. Fund. Math. 1 (1920), 11-16.

[13] R. Sikorski: Boolean Algebras. Ergebnisse der Math. (new series) 25, 2nd ed., Springer Verlag, Berlin 1964.

[14] A. Tarski: Zur Grundlegung der Boolescher Algebra. Fund. Math. 24 (1935), 177-198.

[15] A. and C. Ionescu Tulcea: Topics in the theory of lifting. Ergebnisse der Math. 48, Springer Verlag, Berlin 1969.

[16] H. E. White, Jr.: An example involving Baire spaces. Proc. Amer. Math. Soc. 48 (1975), 228-230.

ON RINGS OF CONTINUOUS FUNCTIONS

Dedicated to Professor K. Morita, on his sixtieth birthday

Jun-iti Nagata
Amsterdam

In the following discussions all topological spaces are at least Tychonoff, and all mappings are continuous. $C(X)$ ($C^*(X)$) denotes the ring of all real-valued continuous functions (real-valued bounded continuous functions) on a Tychonoff space X.

As pointed out by late Professor Tamano, a remarkable property of rings of continuous functions is that they have infinite operations like infinite sum, infinite join etc., and thus it is desirable to study them together with infinite operations. For example, one cannot characterize very important topological properties like metrizability or paracompactness of X in terms of $C(X)$ or $C^*(X)$ as long as they are regarded as ordinary rings with finite operations, but one can give nice characterizations of those properties once infinite operations are taken into consideration. From this point of view the author [7] characterized metrizability and paracompactness in terms of $C(X)$ with operations $\cup$ and $\cap$ for infinitely many elements. H. Tamano [12], Z. Frolík [3] and J. Guthrie [4] also got interesting characterizations of paracompact spaces, Čech complete spaces and other spaces in terms of $C(X)$ and $C^*(X)$ though they did not necessarily aim characterizations by purely <u>internal</u> properties of $C(X)$ or $C^*(X)$. The purpose of this paper is to extend characterization to some generalizations of metric spaces and also to discuss relations between $C^*(X)$ and uniformities of X.

<u>Remark</u>. Only $C^*(X)$ will be used in the following though many results can be extended to $C(X)$ with no or slight modification of their forms. For a (not necessarily finite) subset $\{f_\alpha \mid \alpha \in A\}$ of $C^*(X)$ $\bigcap_\alpha f_\alpha$ and $\bigcup_\alpha f_\alpha$ are defined as usual; namely
$(\bigcap_\alpha f_\alpha)(x) = \inf\{f_\alpha(x) \mid \alpha \in A\}$ $(\bigcup_\alpha f_\alpha)(x) = \sup\{f_\alpha(x) \mid \alpha \in A\}$
In those theorems where $\bigcup_\alpha f_\alpha$ (or $\bigcap_\alpha f_\alpha$) is involved, it is implied that $\bigcup_\alpha f_\alpha$ (or $\bigcap_\alpha f_\alpha$) is bounded and continuous; also note that N, Q and R denote the natural numbers, the rational numbers and the real numbers, respectively.

As for standard symbols and terminologies of general topology, see [10].

Definition 1: A subset L_0 of $C^*(X)$ is called normal if $\bigcap_\alpha f_\alpha$ and $\bigcup_\alpha f_\alpha$ belong to L_0 for every subset $\{f_\alpha \mid \alpha \in A\}$ of L_0. A sequence $L_1, L_2, \ldots$ of normal subsets of $C^*(X)$ is called a normal sequence. A subset L of $C^*(X)$ is σ-normally generated by the normal sequence $\{L_i \mid i = 1,2,\ldots\}$ if $L = \{f \in C^*(X) \mid$ for every $\varepsilon > 0$ there are subsets $\{f_\beta \mid \beta \in B\}$ and $\{f_\gamma \mid \gamma \in C\}$ of $\bigcup_{i=1}^{\infty} L_i$ such that $\| \bigcap_\beta f_\beta - f \| < \varepsilon$ and $\| \bigcup_\gamma f_\gamma - f \| < \varepsilon \}$. (We may simply say that L is generated by $\{L_i\}$ when the latter is known to be a normal sequence.

In the following is a slight modification of an old theorem proved in [7].

Theorem 0. A Tychonoff space X is metrizable iff $C^*(X)$ is σ-normally generated by a normal sequence.

Proof. The "if" part of this theorem is implied by Corollary 8 of [7]. The proof of "only if" part is also not so difficult if we put $L_n = \{f \in C^*(X) \mid \|f\| \leqq n, |f(x) - f(y)| \leqq n \rho(x,y)$ for all $x, y \in X\}$. Some works are necessary to choose, for given $f \in C^*(X)$ and $\varepsilon > 0$, a subset $\{f_\beta \mid \beta \in B\}$ of $\bigcup L_n$ such that $\bigcap_\beta f_\beta \in C^*(X)$ and such that $\| \bigcap_\beta f_\beta - f \| < \varepsilon$, but the detail is left to the reader. (In view of Corollary 8 of [7] we know that a weaker condition is sufficient for the metrizability of X.
The author, however, needs the stronger condition for L as given in Definition 1 to characterize other spaces in the following, and he does not know if the condition can be weakened there or not.)

Among the various generalizations of metric spaces which are actively being studied M-space due to K. Morita [5] and p-space due to A.V. Archangelskii [1] are some of the most important ones. M and p coincide and are especially good if combined together with paracompactness. In fact,

Theorem (K. Morita - A.V. Archangelskii). The following conditions for X are equivalent:
(1) X is paracompact and M,
(2) X is paracompact and p,
(3) X is the pre-image of a metric space by a perfect mapping.

Thus our first aim is to characterize paracompact M-spaces in terms of $C^*(X)$.

<u>Definition</u>. A maximal ideal J in $C^*(X)$ is called <u>fixed</u> iff for every subset $\{f_\alpha \mid \alpha \in A\}$ of J such that $\bigcup_\alpha f_\alpha \in C^*(X)$, $\bigcup_\alpha f_\alpha \in J$ holds. A subset K of $C^*(X)$ is <u>fixed</u> iff there is a fixed maximal ideal which contains K; otherwise K is called <u>free</u>. A subset H of $C^*(X)$ is called <u>strongly free</u> iff there is a subset $\{f_\beta \mid \beta \in B\}$ of H such that $\bigcup_\beta f_\beta \in C^*(X)$ and $\bigcup_\beta f_\beta \geq \varepsilon$ for some positive number ε.

<u>Remark</u>. It is easy to see that $K \in C^*(X)$ is fixed iff there is $x \in X$ for which $f(x) = 0$ for all $f \in K$.

The following theorem suggests us what form of theorem we can expect to characterize paracompact M-spaces.

<u>Theorem 1.</u> <u>Let</u> f <u>be a map from</u> X <u>onto</u> Y. <u>Then</u> f <u>induces an imbedding of</u> $C^*(Y)$ <u>into</u> $C^*(X)$ <u>if</u> $g \in C^*(Y)$ <u>is associated with</u> $g \circ f \in C^*(X)$. <u>Then</u> f <u>is a perfect map iff the induced imbedding is such that for every free maximal ideal</u> J <u>in</u> $C^*(X)$, $J \cap C^*(Y)$ <u>is free in</u> $C^*(X)$.

To prove this theorem we need the following lemma whose proof is left to the reader.

<u>Lemma 1.</u> <u>Let</u> f <u>be a map from</u> X <u>onto</u> Y. <u>Then</u> f <u>is a perfect map iff for every free (= has no cluster point) maximal</u> z-<u>filter</u> (<u> = filter consisting of zero sets where we mean by a zero set the set of all zeros of a real-valued continuous function)</u> $\mathcal{F}$ <u>in</u> X, $f(\mathcal{F}) = \{f(F) \mid F \in \mathcal{F}\}$ <u>is free in</u> Y.

<u>Proof of Theorem 1.</u> The first half of the claim is obvious, so only the last half will be proved.

Assume that f is a perfect map and J is a given free maximal ideal in $C^*(X)$. For each $\phi \in C^*(X)$ and $\varepsilon > 0$, we put $Z_\varepsilon(\phi) = \{x \mid |\phi(x)| \leq \varepsilon\}$.
(This symbol will be used throughout the rest of the paper).
Further, let
$\mathcal{F}(J) = \{Z \mid Z$ is a zero set in X which contains $Z_\varepsilon(\phi)$ for some $\phi \in J$ and for some $\varepsilon > 0\}$.
Then $\mathcal{F}(J)$ is obviously a free z-filter.
Expand $\mathcal{F}(J)$ to a maximal z-filter $\mathcal{F}_0$. Then since f is perfect,

by Lemma 1 $f(\mathcal{F}_0)$ is free in Y. Let x be an arbitrary point of X, and let $f(x) = y$. Then there is $Z \in \mathcal{F}_0$ such that $y \notin f(Z)$.
Since f(Z) is a closed set, there is $\phi \in C^*(Y)$ such that

$$\phi(y) > 0, \ \phi(u) = 0 \text{ for all } u \in f(Z).$$

Then $\phi \circ f(x) > 0$, and $\phi \circ f \in C^*(Y)$, where $C^*(Y)$ is considered to be imbedded in $C^*(X)$.
To prove $\phi \circ f \in J$, let $\phi \circ f = \psi$. Then $J' = C^*(X)\psi + J$ is an ideal of $C^*(X)$ containing J. For each $\xi \in J$, and $\varepsilon > 0$, $Z_\varepsilon(\xi) \cap$ $\cap Z \neq \emptyset$, because these sets both belong to $\mathcal{F}_0$.
Since $\psi(Z) = 0$, this implies $|\alpha\psi + \xi| \leq \varepsilon$ for every $\alpha \in C^*(X)$ and at some point of X. Thus $J' \neq C^*(X)$, which implies $J' = J$ because J is maximal. Thus $\psi \in J$. Namely $\psi \in J \cap C^*(Y)$. Hence $J \cap C^*(Y)$ is free in $C^*(X)$.

Conversely, to prove the "if" part of the theorem, let $\mathcal{F}$ be a free maximal z-filter in X. Put

$$J = \{\psi \in C^*(X) \mid Z_\varepsilon(\psi) \in \mathcal{F} \text{ for all } \varepsilon > 0\}.$$

Then J is a free maximal ideal in $C^*(X)$. To see that J is maximal, let J' be an ideal such that $J \subsetneqq J'$. Select $\phi \in J' - J$; then $Z_\varepsilon(\phi) \notin \mathcal{F}$ for some $\varepsilon > 0$. Since $\mathcal{F}$ is maximal, this implies $Z_\varepsilon(\phi) \cap Z = \emptyset$ for some $Z \in \mathcal{F}$. Put $\psi = \min(0, |\phi| - \varepsilon)$; then $\psi \in J$, because $Z_\sigma(\psi) \supset Z$ for all $\sigma > 0$. Thus $\phi^2 + \psi^2 \in J'$ and $\phi^2 + \psi^2 \geq \frac{\varepsilon^2}{4}$, which imply $J' = C^*(X)$. Therefore J is maximal.
Now, we claim that $f(\mathcal{F})$ has no cluster point in Y. To see it, let $y \in Y$ be arbitrary and select $x \in f^{-1}(y)$. Since by the condition of the theorem $J \cap C^*(Y)$ is free in $C^*(X)$, there is $\phi \in C^*(Y)$ such that $\phi \circ f \in J$ and $\phi \circ f(x) > 0$. Let $\phi \circ f(x) = \varepsilon$. Then $Z_{\frac{\varepsilon}{2}}(\phi \circ f) \cap f^{-1}(y) = \emptyset$, which implies $y \notin f(Z_{\frac{\varepsilon}{2}}(\phi \circ f))$. On the other hand $Z_{\frac{\varepsilon}{2}}(\phi \circ f) \in \mathcal{F}$ follows from the definition of J. Since $f(Z_{\frac{\varepsilon}{2}}(\phi \circ f)) = Z_{\frac{\varepsilon}{2}}(\phi)$, $|\phi(u)| \leq \frac{\varepsilon}{2}$ holds for all $u \in f(Z_{\frac{\varepsilon}{2}}(\phi \circ f))$. Let $V = \{u \in Y \mid \phi(u) > \frac{\varepsilon}{2}\}$; then V is an open nbd of y which is disjoint from $f(Z_{\frac{\varepsilon}{2}}(\phi \circ f))$. Thus y is no cluster point of $f(\mathcal{F})$; namely $f(\mathcal{F})$ is free. Hence by Lemma 1, f is a perfect map.
Now, we can characterize paracompact M-spaces in terms of $C^*(X)$ as follows.

Theorem 2. A Tychonoff space X is paracompact and M iff there is a σ-normally generated subring L of $C^*(X)$ such that for every free maximal ideal J in $C^*(X)$, $J \cap L$ is free.

Proof of the "only if" part. Let X be paracompact and M; then by the previously mentioned Morita-Archangelskii's theorem there is a perfect map from X onto a metric space Y. By Theorem 1 this map induces an imbedding $C^*(Y) \cong L \subset C^*(X)$ satisfying the condition of this theorem. It easily follows from Theorem 0 that L is σ-normally generated in $C^*(X)$.

To prove the "if" part we need some lemmas.

Lemma 2. *Let $\mathcal{V} = \{V_\alpha \mid \alpha < \tau\}$ be a well-ordered open cover of X such that $V_\alpha = \{x \mid f_\alpha(x) > 0\} \cap \{x \mid g_\alpha(x) > 0\}$, $\alpha < \tau$, where f_α, $g_\alpha \in C^*(X)$ for all $\alpha < \tau$, (α and τ denote ordinal numbers). If $\bigcup_{\beta \in B} f_\beta$ and $\bigcup_{\beta \in B} g_\beta$ belong to $C^*(X)$ for every subset B of $\{\alpha \mid 0 \leq \alpha < \tau\}$, then $\mathcal{V}$ has a σ-discrete open refinement consisting of cozero open sets (= complements of zero sets).*

Proof. Note that $V_\alpha = \{x \mid h_\alpha(x) > 0\}$ for $h_\alpha = f_\alpha \cap g_\alpha$ and that $\bigcup_{\alpha < \beta} h_\alpha \in C^*(X)$ for every $\beta \leq \tau$ easily follows from the assumption of lemma. Let

$$V_{1\alpha} = \{x \mid h_\alpha(x) > \tfrac{1}{2}\},$$

$$V_{n\alpha} = \{x \mid h_\alpha(x) > \tfrac{1}{2} - \tfrac{1}{2^2} - \dots \tfrac{1}{2^n}\} \qquad n = 2,3,\dots .$$

Then $V_{n\alpha} \subset V_{n+1\alpha} \subset V_\alpha$. Further, let

$$W_{n1} = V_{n1},$$

$$W_{n\alpha} = \{x \mid x \in V_{n\alpha},\ \bigcup_{\beta < \alpha} h_\beta(x) < \tfrac{1}{2} - \tfrac{1}{2^2} - \dots - \tfrac{1}{2^{n+1}}\}.$$

Then each $W_{n\alpha}$ is a cozero set. It is also obvious that $\{W_{n\alpha} \mid n = 1,2,\dots;\ \alpha < \tau\}$ covers X. Since $W_{n\alpha} \subset V_\alpha$, this cover refines $\mathcal{V}$. Thus it suffices to show that $\{W_{n\alpha} \mid \alpha < \tau\}$ is discrete for each fixed n.

Let $x \in X$ satisfy $x \in V_{n+1\alpha}$ and $x \notin V_{n+1\beta}$ for all $\beta < \alpha$, where $\alpha \leq \tau$. Then $h_\beta(x) \leq \frac{1}{2} - \frac{1}{2^2} - \dots - \frac{1}{2^n} - \frac{1}{2^{n+1}}$, $\beta < \alpha$.

Thus $\bigcup_{\beta < \alpha} h_\beta(x) \leq \frac{1}{2} - \frac{1}{2^2} - \dots - \frac{1}{2^{n+1}}$, and hence x has a nbd W on which $\bigcup_{\beta < \alpha} h_\beta(x') < \frac{1}{2} - \frac{1}{2^2} - \dots - \frac{1}{2^n}$ holds.

Hence $W \cap W_{n\beta} = \emptyset$ for all $\beta < \alpha$.

On the other hand if $\gamma > \alpha$, then $V_{n+1\alpha}$ is a nbd of x which is disjoint from $W_{n\gamma}$. Therefore $\{W_{n\alpha} \mid \alpha < \tau\}$ is discrete.

<u>Lemma 3. Every open cover $\mathcal{V}$ satisfying the condition of Lemma 2 is normal; namely there is a sequence $\mathcal{V}_1, \mathcal{V}_2, \dots$ of open covers such that</u> $\mathcal{V} > \mathcal{V}_1^* > \mathcal{V}_1 > \mathcal{V}_2^* > \dots$.

<u>Proof</u>. The proof directly follows from Lemma 2 and a known theorem (Proposition D) on page 254 of [10]).

<u>Proof of the "if" part of Theorem 2.</u> First of all we define some notations. Assume that L is (σ-normally) generated by the normal sequence $L_1, L_2, \dots$. Then

$$L_m^+ = \{f^+ \mid f \in L_m\}, \text{ where } f^+ = f \cup 0,$$

$$L_m^- = \{f^- \mid f \in L_m\}, \text{ where } f^- = f \cap 0,$$

$$K_m = L_m^+ \cup (-L_m^-).$$

For $x \in X$ and $n, m \in N$,

$$A_n^m(x) = \{f \mid f \in K_m, f(x) > \frac{2}{n}\},$$

$$U_n^m(x) = \{y \mid \cap \{f(y) \mid f \in A_n^m(x)\} > \frac{1}{3n} \quad .$$

$$V_n^m(x) = \{y \mid \cap \{f(y) \mid f \in A_n^m(x)\} > \frac{1}{2n} \quad .$$

$$W_n^m(x) = \{y \mid \cap \{f(y) \mid f \in A_n^m(x)\} > \frac{1}{n} \quad .$$

Then $U_n^m(x)$, $V_m^m(x)$ and $W_n^m(x)$ are open sets satisfying

$$W_n^m(x) \subset V_n^m(x) \subset U_n^m(x),$$

because $\cap \{f \mid f \in A_n^m(x)\}$ is continuous.

The proof will be carried out in several steps.

<u>Claim 1.</u> Let $\mathcal{F}$ be a free maximal z-filter in X. Then for each $x \in X$, there are $n, m \in N$ and $Z \in \mathcal{F}$ such that $U_n^m(x) \cap Z = \emptyset$. To prove it, let

$$J = \{\psi \in C^*(X) \mid Z_\varepsilon(\psi) \in \mathcal{F} \quad \text{for all} \quad \varepsilon > 0\}.$$

Then as proved before for Theorem 1, J is a free maximal ideal in $C^*(X)$. Hence $J \cap L$ is free by the condition of the present theorem. Namely there is $f_0 \in J \cap L$ such that $f_0(x) > 0$ or $-f_0(x) > 0$. Assume that the former is true; then there is $f \in \bigcup_{m=1}^{\infty} L_m$ such that $f(x) > 0$, $f \leq f_0$ because L is generated by $\{L_m\}$. (Recall Definition). Thus there are $n, m \in N$ for which $f \in L_m$, $f^+ \in L_m^+$ and $f^+(x) > \frac{2}{n}$. Then $Z_\sigma(f_0) \subset Z_\sigma(f_0^+) \subset Z_\sigma(f^+)$, for every $\sigma > 0$. Since $f_0 \in J$, $Z_\sigma(f_0) \in \mathcal{F}$ and accordingly $Z_\sigma(f^+) \in \mathcal{F}$ for every $\sigma > 0$, which implies $f^+ \in J$. On the other hand $f^+ \in A_n^m(x)$ follows from the above observation on f^+. Hence $U_n^m(x) \subset \{y \mid f^+(y) > \frac{1}{3n}\}$. This implies $U_n^m(x) \cap Z_{\frac{1}{3m}}(f^+) = \emptyset$. Since $Z_{\frac{1}{3m}}(f^+) \in \mathcal{F}$, our claim is proved. Even if

$-f_0(x) > 0$ is assumed, we can prove our claim in a similar way.

<u>Claim 2.</u> Let $Y \subset X$ to define

$$M_n^m(Y) = \mathrm{Int}\,[\cap\{U_n^m(x) \mid x \in Y\}] \cap \mathrm{Int}\,[\cap\{X - W_n^m(x) \mid x \in X - Y\}],$$

$\mathcal{M}_n^m = \{M_n^m(X) \mid Y \subset X\}$, where $m, n \in N$. Then each $\mathcal{M}_n^m$ is a normal open cover of X.

To prove it, define for m, $n \in N$ and $x' \in X$,

$$P_n^m(x') = \{y \mid \cup\{f(y) \mid f \in K_m,\ f(x') < \frac{2}{3n}\} < \frac{1}{n}\} \cap \{y \mid \cap\{f(y) \mid f \in K_m,$$

$$f(x') > \frac{1}{2n}\} > \frac{1}{3n}\}.$$

Furthermore, define $\mathcal{P}_n^m = \{P_n^m(x') \mid x' \in X\}$.

Then by Lemma 3, $\mathcal{P}_n^m$ is a normal open cover, because each $P_n^m(x')$ satisfies the condition of V_α in Lemma 2, since L_m is a normal set. For each $x' \in X$, let $Y = \{x \mid \cap\{f(x') \mid f \in A_n^m(x)\} > \frac{1}{2n}\}$.

Then it is not difficult to prove that $P_n^m(x') \subset M_n^m(Y)$. Thus $\mathcal{P}_n^m <$ $< \mathcal{M}_n^m$, and hence $\mathcal{M}_n^m$ is also a normal open cover.

<u>Claim 3.</u> $S(x, \mathcal{M}_n^m) \subset U_n^m(x)$ at each point x of X.

To prove it, let $M_n^m(Y)$ be an arbitrary element of $\mathcal{M}_n^m$ which contains x. Then it follows from the definition of $M_n^m(Y)$ that $x \in Y$.

Thus the same definition implies $M_n^m(Y) \subset U_n^m(x)$. Therefore $S(x, \mathcal{M}_n^m) \subset$ $\subset U_n^m(x)$.

Now, we are in a position to complete our proof. Combine claim 1 and claim 3; then we see that for every free maximal z-filter $\mathcal{F}$ and for each $x \in X$ there are m, $n \in N$ and $Z \in \mathcal{F}$ such that $S(x, \mathcal{M}_n^m) \cap Z =$ $= \emptyset$. Since each $\mathcal{M}_n^m$ is normal by claim 2, there is a sequence $\mathcal{U}_1$, $\mathcal{U}_2, \ldots$ of open covers of X such that for each (m,n) and for some i, $\mathcal{U}_i < \mathcal{M}_n^m$ and such that $\mathcal{U}_1 > \mathcal{U}_2^* > \mathcal{U}_2 > \mathcal{U}_3^* > \ldots$. Then for every free maximal z-filter $\mathcal{F}$ and for each $x \in X$, there are some i and some $Z \in \mathcal{F}$ such that $S(x, \mathcal{U}_i) \cap Z = \emptyset$. Assume that $F_1 \supset F_2 \supset \ldots$ is a decreasing sequence of nonempty closed sets in X such that for a fixed point x, $S(x, \mathcal{U}_i) \supset F_k$ holds for each i and for some k. Let $\mathcal{F}$ be a maximal z-filter which is obtained by expanding the collection $\{Z \mid Z$ is a zero set containing F_k for some $K\}$. Then $S(x, \mathcal{U}_i) \cap Z \neq \emptyset$ for every i and every $Z \in \mathcal{F}$. Hence by the above observation we know that $\mathcal{F}$ converges. Since $\cap\{F \mid F \in \mathcal{F}\} \subset \bigcap_{k=1}^\infty F_k$ follows from complete regularity of X, we have $\bigcap_{k=1}^\infty F_k \neq \emptyset$, which proves that X is an M-space.

Let $C_x = \bigcap_{i=1}^\infty S(x, \mathcal{U}_i)$; then as shown in [5], there is a closed map g from X onto a metric space Y such that for each $y \in Y$, $g^{-1}(y) =$

$= C_x$ for some $x \in X$. To prove compactness of the closed set C_x, let $\mathfrak{F}_0$ be a collection of closed subsets of C_x with finite intersection property. Let $\mathfrak{F}'$ be a maximal z-filter which is obtained by expanding the collection $\{Z \mid Z$ is a zero set containing some element of $\mathfrak{F}_0\}$. Then obviously $S(x, \mathcal{U}_i) \cap Z \neq \emptyset$ for every i and every $Z \in \mathfrak{F}'$. Thus $\mathfrak{F}'$ converges, and hence $\bigcap\{F \mid F \in \mathfrak{F}_0\} \neq \emptyset$. Therefore C_x is compact, i.e. g is a perfect map. This proves that X is paracompact, and now the proof of Theorem 2 is complete.

Now, let us turn to another generalization of (complete) metric spaces. Paracompact, Čech complete spaces are characterized as follows.

Theorem (Z. Frolík [21]). *X is a paracompact, Čech complete space iff it is the pre-image of a complete metric space by a perfect map.*

This theorem is in its appearance similar to the previously mentioned Morita-Archangelskii's theorem and indicates that all paracompact Čech complete spaces are paracompact M. In fact the latter theorem is a sort of generalization of the former. Thus it is natural to try to characterize paracompact Čech complete spaces in a similar way as we did for paracompact M-spaces. As a result we obtain the following theorem.

Theorem 3. *A Tychonoff space X is paracompact and Čech complete iff there is a normal sequence $L_1, L_2, \ldots$ of subsets of $C^*(X)$ such that for every free maximal ideal J in $C^*(X)$, $J \cap L_n$ is strongly free for some n.*

Proof. To prove the "only if" part, let X be paracompact and Čech complete. Then by Frolík's theorem there is a perfect map f from X onto a complete metric space Y. Let

$L_n = \{\phi \circ f \mid \phi \in C^*(Y),\ \|\phi\| \leq n.\ |\phi(y) - \phi(z)| \leq n\rho(y,z)$ for every $y, z \in Y\}$,

where we assume ρ is a metric of Y such that $\rho \leq 1$.

Then each L_n is a normal subset of $C^*(X)$.

Let $\mathfrak{F}$ be a maximal z-filter in X which contains $Z_\varepsilon(\psi)$ for all $\psi \in J$ and for all $\varepsilon > 0$. Since J is free, so is $\mathfrak{F}$. Since f is perfect, by Lemma 1 $f(\mathfrak{F})$ is free. Since Y is a complete metric space, there is $\varepsilon > 0$ such that $S_\varepsilon(y) \not\supset f(Z)$ for all $y \in Y$ and for all $Z \in \mathfrak{F}$. For each $y \in Y$ define $\phi_y \in C^*(Y)$ by

$$\phi_y(z) = \rho(z, Y - S_\varepsilon(y)).$$

Let $\psi_y = \phi_y \circ f$; then $\psi_y \in L_1$. Since $Z(\psi_y) \cap Z_\varepsilon(\xi) \neq \emptyset$ for all $\varepsilon > 0$ and all $\xi \in J$, $\psi_y \in J$ follows from maximality of J, where

$$Z(\psi_y) = \{ x \in X \mid \psi_y(x) = 0 \}.$$

(See the proof of Theorem 1.) Thus $\psi_y \in J \cap L_1$ for every $y \in Y$. On the other hand $\bigcup_{y \in Y} \psi_y \geq \varepsilon$ is obvious, and hence $J \cap L_1$ is strongly free.

To prove the "if" part, first note that by Theorem 2, X is at least paracompact and M. Thus it suffices to prove that X is Čech complete. For each $x \in X$ and $n, m \in N$ we define $A_n^m(x)$ and $U_n^m(x)$ exactly in the same way as in the proof of Theorem 2. Now, let $\mathcal{F}$ be an arbitrary free maximal z-filter in X; then we shall prove that there are $n, m \in N$ such that $X - U_n^m(x) \in \mathcal{F}$ for all $x \in X$. This would prove Čech completeness of X by N.A. Shanin's theorem [11]: X is Čech complete iff there is a sequence $\{\mathcal{G}_i \mid i = 1,2,\ldots\}$ of collections of zero-sets with finite intersection property such that i) $\bigcap \{ G \mid G \in \mathcal{G}_i \} = \emptyset$, ii) for every free maximal z-filter $\mathcal{F}$, there is i for which $\mathcal{G}_i \subset \subset \mathcal{F}$. For this end, let

$$J = \{ \psi \in C^*(x) \mid Z_\varepsilon(\psi) \in \mathcal{F} \quad \text{for all} \quad \varepsilon > 0 \}.$$

Then as proved before for Theorem 1, J is a free maximal ideal. Hence $J \cap L_m$ is strongly free for some m. Namely there is a subset $\{ \phi_\alpha \mid \alpha \in A \}$ of $J \cap L_m$ such that $\bigcup_{\alpha \in A} \phi_\alpha \geq \varepsilon$ for some positive number ε. Choose $n \in N$ for which $\frac{\varepsilon}{2} > \frac{2}{n}$. Then for each $x \in X$ there is $\alpha \in A$ such that $\phi_\alpha(x) \geq \frac{\varepsilon}{2} \; \frac{2}{n}$. Thus $\phi_\alpha^+ \in A_n^m(x)$.

Since $0 \leq \phi_\alpha^+ \leq \frac{1}{3n}$ on $Z_{\frac{1}{3n}}(\phi_\alpha)$,

$$U_n^m(x) \subset \{ y \mid \phi_\alpha^+(y) > \tfrac{1}{3n} \} \subset X - Z_{\frac{1}{3n}}(\phi_\alpha).$$

Thus $X - U_n^m(x) \supset Z_{\frac{1}{3n}}(\phi_\alpha) \in \mathcal{F}$. (Note that $\phi_\alpha \in J$).

This proves that $X - U_n^m(x) \in \mathcal{F}$ for every x, and accordingly X is Čech complete.

Next, let us turn to a class of generalized metric spaces which contains all paracompact M-spaces as a proper subclass.

<u>Definition 3</u>. A Tychonoff space X is called a G_δ <u>-space</u> iff it is homeomorphic to a G_δ-set in the product of a metric space and a compact T_2-space.

G_δ-space was defined in [9] as a natural generalization of pa-

racompact M-spaces, because in [8] a paracompact (T_2) M-space was characterized as a closed G_δ-set in the product of a metric space and a compact T_2-space.
In [9] the author gave the following characterizations to G_δ-spaces.

Definition 4. Let f be a continuous map from X onto Y.
Then f is called a complete map if there is a sequence $\mathcal{U}_1, \mathcal{U}_2, \ldots$ of covers of X by cozero sets such that for every free maximal z-filter $\mathcal{F}$ in X satisfying $\mathcal{G}_n = \{X - U \mid U \in \mathcal{U}_n\} \not\subset \mathcal{F}$, $n = 1,2,\ldots$, $f(\mathcal{F})$ is free in Y.

Theorem A. X is a G_δ-space iff it is the pre-image of a metric space by a complete mapping.

Theorem B. X is a G_δ-space iff there are sequences $\{\mathcal{W}_i \mid i = 1, 2,\ldots\}$ and $\{\mathcal{U}_i \mid i = 1,2,\ldots\}$ of open covers of X such that

(1) $\mathcal{U}_1 > \mathcal{U}_2^* > \mathcal{U}_2 > \mathcal{U}_3^* > \ldots$.

(2) if $\mathcal{F}$ is a maximal closed filter such that $F_i \subset W_i \cap S(x, \mathcal{U}_i)$, $i = 1,2,\ldots$ for some $F_i \in \mathcal{F}$, $W_i \in \mathcal{W}_i$ and a fixed point x of X, then $\mathcal{F}$ converges.

Remark. As for Theorem A a somewhat different (and more complicated) form of condition was considered for the map f in [9], but it is easy to prove that the original condition is equivalent with completeness of f as long as X and Y are Tychonoff.

This theorem should be compared with the previously mentioned theorem of Morita-Archangelskii on paracompact M-spaces. Definition 4 should be compared with Lemma 1 to recognize that complete map is a natural generalization of a perfect map. Thus a complete map may be defined more generally for topological spaces X and Y while replacing cozero sets and zero sets in the present definition with open sets and closed sets, respectively.

The following diagram is to clarify relations between generalized metric spaces being discussed in the present paper.

paracompact and Čech complete	$\Longrightarrow$	paracompact and M (or p)	$\Longrightarrow$	G_δ	$\Longrightarrow$	p
$\Vert$		$\Vert$		$\Vert$		
perfect pre-image of a complete metric space		perfect pre-image of a metric space		complete pre-image of a metric space		
$\Vert$		$\Vert$		$\Vert$		
closed G_δ-set in the product of a complete metric space and a compact T_2-space		closed G_δ-set in the product of a metric space and a compact T_2-space		G_δ-set in the product of a metric space and a compact T_2-space.		

It was proved in [9] that an M-space X is a G_δ-space iff it is a p-space but it is not known if the same is true without the assumption that X is an M-space though a negative answer is supposed. Namely

__Problem.__ Give an example of a p-space which is no G_δ-space.
As suggested by Theorems A and B we can easily characterize the G_δ-spaces in terms of $C^*(X)$ in a similar way as we did for two other spaces in Theorems 2 and 3.

__Theorem 4.__ *A Tychonoff space* X *is a* G_δ*-space iff there is a* σ*-normally generated subring* L *of* $C^*(X)$ *and a sequence* $G_1, G_2, \ldots$ *of free subsets of* $C^*(X)$ *such that for every free maximal ideal* J *in* $C^*(X)$ *satisfying* $G_n \not\subset J$, $n = 1,2,\ldots$, $J \cap L$ *is free in* $C^*(X)$.

__Proof.__ The proof is similar to that of Theorem 2, so only a sketch will be given in the following. Let X be a G_δ-space; then by Theorem A there is a metric space Y and a map f from X onto Y, which is complete with respect to open covers $\mathcal{U}_i$, $i = 1,2,\ldots$ of X. Put $G_n = \{\phi \mid \phi \in C^*(X),\ X - Z(\phi) \in \mathcal{U}_n\}$. Then each G_n is a free subset of $C^*(X)$. Now, suppose that J is a given free maximal ideal in $C^*(X)$ such that $G_n \not\subset J$, $n = 1,2,\ldots$. Then let $\mathcal{F}$ be a maximal z-filter containing $Z_\varepsilon(\phi)$ for all $\phi \in J$ and $\varepsilon > 0$. Then we claim that $\mathcal{G}_n = \{X - U \mid U \in \mathcal{U}_n\} \not\subset \mathcal{F}$, $n = 1,2,\ldots$. Since $G_n \not\subset J$, there is $\phi \in G_n - J$. Then $Z(\phi) \in \mathcal{G}_n$. $Z(\phi) \notin \mathcal{F}$ follows from maximality of J, because otherwise $J \subsetneqq C^*(X)\phi + J \neq C^*(X)$ would hold. Thus $\mathcal{F}$ is a free maximal z-filter satisfying $\mathcal{G}_n \not\subset \mathcal{F}$, $n = 1,2,\ldots$. Since f is a complete map, this implies that $f(\mathcal{F})$ is free in Y.
Thus we can use an argument like the one in the proof of Theorem 1

to conclude that $J \cap L$ is free in $C^*(X)$, where L is the isomorphic image of $C^*(Y)$ in $C^*(X)$ induced by the map f. Since L is σ-normally generated, necessity of the condition is proved.

Conversely assume that $C^*(X)$ satisfies the condition of the theorem. To prove that X is a G_δ-space, define a normal sequence $\mathcal{U}_1 > \mathcal{U}_2^* > \mathcal{U}_2 > \mathcal{U}_3^* > \dots$ of open covers of X in the same way as in the last part of the proof of Theorem 2. Further we define $\mathcal{W}_n = \{X - Z_\varepsilon(\psi) \mid \psi \in G_n, \ \varepsilon > 0\}$; then $\mathcal{W}_n$ is an open cover of X. Assume that $\mathcal{F}$ is a given free maximal closed filter in X such that for every n there is $W \in \mathcal{W}_n$ and $F \in \mathcal{F}$ satisfying $F \subset W$. Put $J = \{\psi \mid \psi \in C^*(X), Z_\varepsilon(\psi) \in \mathcal{F}$ for all $\varepsilon > 0\}$. Then as in the proof of Theorem 1, we can prove that J is a free maximal ideal. Moreover we can show that $G_n \not\subset J$, $n = 1,2,\dots$. Because $W = X - Z_\varepsilon(\psi) \supset F \in \mathcal{F}$ for some $\psi \in G_n$ and $\varepsilon > 0$. Hence $Z_\varepsilon(\psi) \notin \mathcal{F}$ proving that $\psi \notin J$. Therefore $J \cap L$ is free in $C^*(X)$. Thus in a similar way as in the proof of Theorem 2 we can prove that for each $x \in X$ there are $Z \in \mathcal{F}$ and i such that $S(x, \mathcal{U}_i) \cap Z = \emptyset$. This means that $F \not\subset S(x, \mathcal{U}_i)$ holds for all $F \in \mathcal{F}$. Hence by Theorem B X is a G_δ-space.

It would be easy to characterize (general) Čech complete spaces and perhaps general M-spaces, too, in terms of $C^*(X)$ by use of a similar method. How about p-spaces? There is another group of generalized metric spaces which can be characterized as <u>images</u> of metric spaces by certain types of maps, e.g. Lašnev space (= closed continuous image of metric space), stratifiable space, σ-space, etc. Is it possible to characterize them by simple properties of $C^*(X)$ as we have done for pre-images of metric spaces? In any case one may need a new technique which is different from the one we have used.

Now, let us turn to an extension of Theorem 0 to another direction. If X is metrizable, then by that theorem $C^*(X)$ is generated by a normal sequence. Then what is the relation between various normal sequences generating $C^*(X)$ and metric uniformities of X ? We shall see in the following that they correspond to each other in certain manner.

<u>Theorem 5</u>. <u>Let</u> X <u>be a metric space; then there is a normal sequence</u> $\{L_n \mid n = 1,2,\dots\}$ <u>generating</u> $C^*(X)$ <u>such that</u> $A(\{L_n\}) = \{f \in C^*(X) \mid$ <u>for every</u> $\varepsilon > 0$ <u>there is</u> $g \in \bigcup_{n=1}^{\infty} L_n$ <u>for which</u> $\|f - g\| < \varepsilon\}$ <u>is equal to the set</u> $V^*(X)$ <u>of all bounded uniformly continuous (real-va-</u>

<u>lued) functions on</u> X. <u>Moreover we can select</u> $\{L_n\}$ <u>satisfying the following condition</u>.

(A) $L_1 \subset L_2 \subset \ldots,$

$f \in L_n$ <u>implies</u> $f \cup 0,\ f \cap 0 \in L_n$,

$f \in L_n$ <u>and</u> $\alpha \in R$ <u>imply</u> $f + \alpha,\ \alpha f \in L_m$

<u>for some</u> $m = m(n, \alpha)$.

<u>Proof</u>. Let $L_n = \{ f \in C^*(X) \mid \|f\| \leq n,\ |f(x) - f(y)| \leq n\, \rho(x,y)$ for all $x,y \in X\}$. Then $\{L_n \mid n = 1,2,\ldots\}$ is a normal sequence satisfying the required conditions.

$A(\{L_n\}) \subset V^*(X)$ is obvious because each element of $\bigcup_{n=1}^{\infty} L_n$ is bounded and uniformly continuous. To prove $V^*(X) \subset A(\{L_n\})$ assume that the metric ρ of X is bounded and also let $f \in V^*(X)$ and $\varepsilon > 0$. Further suppose $\|f\| \leq A$. Select $k \in N$ such that $\rho(x,y) < \frac{1}{k}$ implies $\rho(f(x).f(y)) < \varepsilon$. Then put $F_n = \{x \mid n\varepsilon < f(x) \leq (n+1)\varepsilon\}$, $n = 0, \pm 1, \pm 2, \ldots$. (We define F_n only for such n that satisfies $[n\varepsilon, (n+1)\varepsilon] \cap [-A, A] \neq \emptyset$.

For each $n \in N$, let $p_n \in N$ be such that

$$p_n - 1 < k(A - n\varepsilon) \leq p_n.$$

Put $f_n(x) = p_n \rho(F_n, x) + n\varepsilon$; then $f_n \in L_{q_n}$ for some $q_n \in N$, $f_n(x) = n\varepsilon$ for $x \in F_n$ and $f_n(x) \geq A$ for $x \notin F_{n-1} \cup F_n \cup F_{n+1}$. Thus $(n-1)\varepsilon \leq \bigcap_n f_n(x) \leq n\varepsilon$ holds for each $x \in F_n$.

Therefore $\|f - \bigcap_n f_n\| \leq 2\varepsilon$. Note that $\bigcap_n f_n \in L_m$ for some m (= the largest n for which F_n is defined). This proves $V^*(X) \subset A(\{L_n\})$ and eventual coincidence of these two sets.

<u>Theorem 6</u>. <u>Let</u> $\{L_n \mid n = 1,2,\ldots\}$ <u>be a normal sequence generating</u> $C^*(X)$ <u>and satisfying the condition (A) of the previous theorem. Then there is a metric uniformity (agreeing with the topology) of X for which</u> $V^*(X) = A(\{L_n\})$, <u>where the symbols</u> V^* <u>and</u> A <u>are defined in the same way as in the previous theorem</u>.

<u>Proof</u>. 1. First note that X is metrizable.

For each $x \in X$, $n \in N$ and $v, v' \in Q$ (the rationals) such that $v < v'$, we define

$$B_{nvv'}(x) = \{f \mid f \in L_n,\ f(x) \geq \frac{v + v'}{2}\},$$

$$\tilde{B}_{nvv'}(x) = \{f \mid f \in L_n,\ f(x) \leq \frac{v + v'}{2}\},$$

$U_{nvv'}(x) = \{y \in X \mid \bigcap \{f(y) \mid f \in B_{nvv'}(x)\} > v$ and

$$\bigcup\{f(y) \mid f \in \widetilde{B}_{nvv'}(x)\} < v'\},$$

$$V_{nvv'}(x) = \{y \in X \mid \bigcap\{f(y) \mid f \in B_{nvv'}(x)\} > \frac{5v+v'}{6} \text{ and}$$
$$\bigcup\{f(y) \mid f \in \widetilde{B}_{nvv'}(x)\} < \frac{v+5v'}{6}\},$$

$$S_{nvv'}(x) = \{y \in X \mid \bigcap\{f(y) \mid f \in B_{nvv'}(x)\} > \frac{2v+v'}{3} \text{ and}$$
$$\bigcup\{f(y) \mid f \in \widetilde{B}_{nvv'}(x)\} < \frac{v+2v'}{3}\}.$$

Then $S_{nvv'}(x) \subset V_{nvv'}(x) \subset U_{nvv'}(x)$, and they are all open nbds of x. For any $(n,v,v') \in N \times Q \times Q$ with $v < v'$ and for any $Y \subset X$ we define

$$U_{nvv'}(Y) = \text{Int}\,[\bigcap\{U_{nvv'}(x) \mid x \in Y\} \cap (\bigcap\{X - S_{nvv'}(x) \mid x \in X - Y\})],$$

$$\mathcal{U}_{nvv'} = \{U_{nvv'}(Y) \mid Y \subset X\}.$$

Then $\mathcal{U}_{nvv'}$ is obviously an open cover of X. Furthermore we claim:

(a) For every $(n,v,v') \in N \times Q \times Q$, there are (s,s') and $(t,t') \in Q \times Q$ such that

$$\{U_{nss'}(y) \cap U_{ntt'}(y) \mid y \in X\} < \mathcal{U}_{nvv'}.$$

To see it, let y be a fixed point of X.

Then for each $x \in X$, either $y \in V_{nvv'}(x)$ or $y \notin V_{nvv'}(x)$ holds. If $y \in V_{nvv'}(x)$, then it is easy to see that $U'(y) = U_{nss'}(y) \cap U_{ntt'}(y) \subset U_{nvv'}(x)$, where $s = \frac{v+5v'}{6} - \frac{v'-v}{12}$, $s' = \frac{v+5v'}{6} + \frac{v'-v}{12}$, $t = \frac{5v+v'}{6} - \frac{v'-v}{12}$, $t' = \frac{5v+v'}{6} + \frac{v'-v}{12}$.

If $y \notin V_{nvv'}(x)$, then either $\bigcap\{f(y) \mid f \in B_{nvv'}(x)\} \leq \frac{5v+v'}{6}$ or $\bigcup\{f(y) \mid f \in \widetilde{B}_{nvv'}(x)\} \geq \frac{v+5v'}{6}$.

If the former is the case, then for every $\varepsilon > 0$ there is $f_\varepsilon \in B_{nvv'}(x)$ such that $f_\varepsilon(y) \leq \frac{5v+v'}{6} + \varepsilon$. Let $f = \bigcap_{\varepsilon > 0} f_\varepsilon$; then $f \in \widetilde{B}_{nvv'}(x)$, $f(y) \leq \frac{5v+v'}{6}$. Hence $f \in \widetilde{B}_{ntt'}(y)$, and hence for each $u \in U'(y)$ $f(u) \leq t'$. On the other hand, since $f \in B_{nvv'}(x)$, $f(w) > \frac{2v+v'}{3} > t'$ for each $w \in S_{nvv'}(x)$.

Thus $U'(y) \cap S_{nvv'}(x) = \emptyset$.

Even if the latter is the case, we can prove the same in a similar way. Thus we obtain

$$U'(y) \subset U_{nvv'}(Y), \text{ where}$$

$$Y = \{x \in X \mid y \in V_{nvv'}(x)\}.$$

This proves that $\{U_{nss'}(y) \cap U_{ntt'}(y) \mid y \in x\} < \mathcal{U}_{nvv'}$, as claimed in

(a). On the other hand the following relation is almost obvious:
(b) for every $(n,v,v') \in N\times Q\times Q$ and $x\in X$,

$$S(x, \mathcal{U}_{nvv'}) \subset U_{nvv'}(x).$$

As easily seen, $\{U_{nvv'}(x) \mid (n,v,v') \in N\times Q\times Q,\ v<v'\}$ forms a nbd base at each $x\in X$, and therefore (b) implies that $\{S(x, \mathcal{U}_{nvv'}) \mid (n,v,v')\in N\times Q\times Q,\ v<v'\}$ is also a nbd base at x.

Now we can conclude this section of the proof with the following observation:

$$\mu = \{ \bigwedge_{i=1}^{\ell} \mathcal{U}_{n_i v_i v_i'} \mid (n_i,v_i,v_i') \in N\times Q\times Q,\ v_i < v_i',\ i = 1\dots\ell;\ \ell = 1,2,\dots\}$$

is a countable (= metric) uniformity base agreeing with the topology of X.

Let $\mathcal{U}_{nvv'} \in \mu$ be given; then there are (s,s') and (t,t') satisfying (a). Put $\mathcal{U} = \mathcal{U}_{nss'} \wedge \mathcal{U}_{ntt'}$; then by (b) $\mathcal{U}^{\Delta} < \mathcal{U}_{nvv'}$ while $\mathcal{U} \in \mu$. This proves that μ is a uniformity base while we have seen before that this uniformity agrees with the topology of X.

2. From now on we regard X as a (metrizable) uniform space with the uniformity defined by μ.
The objective of the present section is to prove that $A(\{L_n\}) \subset V^*(X)$. It suffices to show that every $f \in \bigcup_{n=1}^{\infty} L_n$ is uniformly continuous with respect to μ. Assume $f\in L_n$, $a\leq f\leq b$ and $a,b\in Q$. Given $\varepsilon > 0$; then choose $k\in N$ for which $\frac{b-a}{k} < \varepsilon$. Put $a_i = a + \frac{b-a}{k}$ $i = -1,0,1,\dots,k,k+1$. Assume that x and y are points of X satisfying

$$y \in U_{na_{-1}a_1}(x) \cap U_{na_0a_2}(x) \cap U_{na_1a_3}(x) \cap \dots \cap U_{na_{k-2}a_k}(x) \cap U_{na_{k-1}a_{k+1}}(x)$$

Then assume that $a_i \leq f(x) < a_{i+1}$; then

$$f\in B_{na_{i-1}a_{i+1}}(x) \cap \widetilde{B}_{na_ia_{i+2}}(x),\ \text{and hence}$$

$$U_{na_{i-1}a_{i+1}}(x) \subset \{y \mid f(y) > a_{i-1}\},\ \text{i.e.}\ f(y) > a_{i-1}.$$

Similarly we can show $f(y) < a_{i+2}$.
Thus $|f(x) - f(y)| < 2\varepsilon$ proving that f is uniformly continuous.

3. Finally we are going to prove $V^*(x) \subset A(\{L_n\})$.
Note that condition (A) will be fully used for the first time in thi section. Let us begin with simple remarks, of which only the last on is given a proof.

(i) Let $v_1 < v_2 < v_3 < v_4$ be rationals satisfying

$\frac{v_1 + v_4}{2} = \frac{v_2 + v_3}{2}$; then $U_{nv_2v_3}(x) \subset U_{nv_1v_4}(x)$ for every $n \in N$ and $x \in X$.

(ii) Let $m < n$ be natural numbers; then

$U_{mvv'}(x) \subset U_{nvv'}(x)$ for every $v, v' \in Q$ and $x \in X$.

(iii) Let $v, v', \alpha \in Q$ and $n \in N$; then there is $m \in N$ (independent from x) such that

$U_{m,v-\alpha,v'-\alpha}(x) \subset U_{nvv'}(x)$ for all $x \in X$.

To see it, let $m = m(n,-\alpha)$ in the condition (A), i.e. $f \in L_n$ implies $f - \alpha \in L_m$. Let $y \in U_{m,v-\alpha,v'-\alpha}(x)$; then

$\bigcap\{f(y) \mid f \in B_{m,v-\alpha,v'-\alpha}(x)\} \geq v - \alpha + \varepsilon > v - \alpha$ and
$\bigcup\{f(y) \mid f \in \tilde{B}_{m,v-\alpha,v'-\alpha}(x)\} \leq v' - \alpha - \varepsilon < v' - \alpha$ for some $\varepsilon > 0$.

Let $f \in B_{nvv'}(x)$; then

$f - \alpha \in B_{m,v-\alpha,v'-\alpha}(x)$, and hence

$f(y) - \alpha \geq v - \alpha + \varepsilon$, i.e. $f(y) \geq v + \varepsilon$.

Thus $\bigcap\{f(y) \mid f \in B_{nvv'}(x)\} \geq v + \varepsilon > v$.

Similarly $\bigcup\{f(y) \mid f \in \tilde{B}_{nvv'}(x)\} \leq v' - \varepsilon < v'$.

Hence $y \in U_{nvv'}(x)$, proving $U_{m,v-\alpha,v'-\alpha}(x) \subset U_{nvv'}(x)$.

Combining (i),(ii) and (iii) we can conclude that for every $\mathcal{U} \in \mathcal{M}$, there is $(n,v,v') \in N \times Q \times Q$ such that $U_{nvv'}(x) \subset S(x,\mathcal{U})$ for all $x \in X$.

Now we are in a position to prove that for every $f \in V^*(x)$ and for every $\varepsilon > 0$, there is $\psi \in \bigcup_{n=1}^{\infty} L_n$ such that $\|f - \psi\| < \varepsilon$. Assume $\|f\| \leq K$. Since f is uniformly continuous, there is $(m,v,v') \in N \times Q \times Q$ for which

$\{U_{mvv'}(x) \mid x \in X\} < \{f^{-1}((n\varepsilon,(n+2)\varepsilon)) \mid n = 0, \pm 1, \dots\}$.

Let x be a given point of X, and suppose that $n\varepsilon \leq f(x) < (n+1)\varepsilon$. Then

$U_{mvv'}(x) \subset f^{-1}(((n-1)\varepsilon, (n+2)\varepsilon))$.

Thus for each $y \notin f^{-1}(((n-1)\varepsilon, (n+2)\varepsilon))$, $y \notin U_{mvv'}(x)$.

Namely either there is $f_{xy} \in L_m$ satisfying $f_{xy}(x) \leq \frac{v+v'}{2}$, $f_{xy}(y) \geq v'$ or else there is $f_{xy} \in L_m$ satisfying $f_{xy}(x) \geq \frac{v+v'}{2}$, $f_{xy}(y) \leq v$.

Hence there are $\alpha, \beta \in R$ such that $g_{xy} = 0 \cup (\alpha f_{xy} + \beta)$ satisfies $g_{xy} \geq 0$, $g_{xy}(x) = 0$, $g_{xy}(y) \geq K - n\varepsilon$, and $g_{xy} \in L_p$ for some p independent from y. Let $h_{xy} = g_{xy} + n\varepsilon \in L_q$, where q is independent from y. Thus $\Phi_x = \bigcup\{h_{xy} \mid y \notin U_{mvv'}(x)\}$ satisfies

$\Phi_x \in L_q$, $\Phi_x \geq n\varepsilon$, $\Phi_x(x) = n\varepsilon$, $\Phi_x(y) \geq K$ for all $y \notin U_{mvv'}(x)$.

Observe that q may be assumed to be common to all $x \in f^{-1}([n\varepsilon, (n+1)\varepsilon))$. Thus

$\psi_n = \bigcap \{\phi_x \mid x \in f^{-1}([n\varepsilon, (n+1)\varepsilon))\}$ satisfies

$\psi_n \in L_q$, $\psi_n(x) = n\varepsilon$ for all $x \in f^{-1}([n\varepsilon, (n+1)\varepsilon))$,

$\psi_n \geq n\varepsilon$, and $\psi_n(y) \geq K$ for all $y \notin f^{-1}(((n-1)\varepsilon, (n+2)\varepsilon))$.

Finally put $\psi = \bigcap_n \psi_n$; then $\psi \in L_\ell$ for some $\ell \in N$.
Moreover it is easy to see that $\| f - \psi \| \leq 2\varepsilon$.
Therefore $V^*(x) \subset A(\{L_n\})$, which completes the proof of the theorem

As proved in [6] (Lemma 2), $V^*(X)$ determines a metric uniformity of a metrizable space X, and hence Theorems 5 and 6 indicate that normal sequences generating $C^*(X)$ and satisfying (A) and metric uniformities of X are corresponding to each other though the correspondence is not one-to-one, because different normal sequences can induce the same $V^*(X)$.

References

1. A.V. Archangelskii, On a class of spaces containing all metric and all locally bicompact spaces, Soviet Math. 4(1963), 751-754.
2. Z. Frolík, On the topological product of paracompact spaces, Bull. Acad. Polon. Sci. Sér. Sci. Math. Astronom. Phys. 8(1960) 747-750.
3. Z. Frolík, A characterization of topologically complete spaces in the sense of E. Čech in terms of convergence of functions, Czechoslovak Math. J. 13(1963), 148-151.
4. J.A. Guthrie, Metrization and paracompactness in terms of real functions, Bull. Amer. Math. Soc. 80(1974), 720-721.
5. K. Morita, Products of normal spaces with metric spaces, Math. Ann. 154(1964), 365-382.
6. J. Nagata, On lattices of functions on topological spaces and o functions on uniform spaces, Osaka J. Math. 1(1949), 166-181.
7. J. Nagata, On coverings and continuous functions, J. Inst. Poly tech. Osaka City Univ. 7(1956), 29-38.
8. J. Nagata, A note on M-space and topologically complete space, Proc. Japan Acad. 45(1969), 541-543.
9. J. Nagata, On G_δ-sets in the product of a metric space and a compact space I, II, Proc. Japan Acad. 49(1973), 179-186.
10. J. Nagata, Modern General Topology, Amsterdam-London-Groningen, 1974.

11. N.A. Shanin, On the theory of bicompact extensions of topological spaces, Dokl. Akad. Nauk SSSR 38(1943), 154-156.

12. H. Tamano, On rings of real valued continuous functions, Proc. Japan Acad. 34(1958), 361-366.

Combinatorial properties of uniformities

Jan Pelant, Prague (Czechoslovakia)

In [S], A.H.Stone raised a question of whether each uniform space has a basis consisting of locally finite covers (recall the A.Stone theorem asserting that each metric space is paracompact).It is shown easily in [I] that the existence of a basis consisting of locally finite covers is equivalent to the existence of a basis consisting of point-finite covers. Stone's problem is restated in [I] and other related problems are pointed out (e.g. the problem of when the Ginsburg-Isbell derivative forms a uniformity, see $[P_1]$,[PPV]). The negative answer to Stone's problem was given independently by E.Ščepin and myself in 1975. Hence the class of all spaces with a point-finite basis forms a "nice" proper epireflective subcategory of UNIF. However, it appears that even spaces having point-finite bases are very wild and that perhaps the best uniform spaces are those having bases consisting of σ-disjoint covers. (A σ-disjoint basis implies the existence of a point-finite base (see e.g. [RR], $[P_1]$, but the converse is not true, (see $[P_2]$)). This paper illustrates the use of "combinatorial" (or discrete) reasoning, as opposed to "continuous" reasoning, in the theory of uniform spaces. This approach seems particularly applicable to problems dealing with covering properties of uniformities.

We are going to estimate point character of some uniform spaces. Finally, we show that the properties of cardinal reflections in UNIF depends on set-theoretical assumptions.

Notation: Let A be a set and let α be a cardinal. We define:

$\mathcal{P}(A) = \{B \mid B \subset A\}$

$[A]^{<\alpha} = \{B \subset A \mid |B| < \alpha\}$, $[A]^{\leq\alpha} = \{B \subset A \mid |B| \leq \alpha\}$

$[A]^{\alpha} = \{B \subset A \mid |B| = \alpha\}$; the meaning of $[A]^{>\alpha}$ and $[A]^{\geq\alpha}$ is obvious

Definition: Let $\mathcal{A}$ be a collection of sets.

1) The order ($\operatorname{ord} \mathcal{A}$) of $\mathcal{A}$ is defined by $\operatorname{ord} \mathcal{A} = \sup \{ |\mathcal{B}| \mid \mathcal{B} \in [\mathcal{A}]^{<\omega_0}, \bigcap \mathcal{B} \neq \emptyset \}$.

2) The degree ($\deg \mathcal{A}$) of $\mathcal{A}$ is defined by $\deg \mathcal{A} = \max (\omega_0, \sup \{ |\mathcal{B}|^+ \mid \mathcal{B} \subset \mathcal{A}, \bigcap \mathcal{B} \neq \emptyset \})$.

Remark: Following [I], if a uniform space $(X, \mathcal{V})$ has a basis consisting of covers of order at most $(n+1)$, (n is a non-negative integer) and does not have a basis consisting of covers of order at most n, then $(X, \mathcal{V})$ is said to be n-dimensional ($\Delta\delta(X, \mathcal{V}) = n$). If there exists no integer n such that $\Delta\delta(X, \mathcal{V}) \leq n$, then we set $\Delta\delta(X, \mathcal{V}) = \infty$.

Definition: The point-character $pc(X, \mathcal{V})$ of a uniform space $(X, \mathcal{V})$ is defined to be the least cardinal α such that $(X, \mathcal{V})$ has a basis consisting of covers whose degrees are at most α.

Basic notation: Let p be a positive integer and let M be a non-empty set. The symbol $\mathcal{K}^p(M)$ denotes the set of all sequences $\{C_j\}_{j=1}^p$ such that $C_p \subset M$ and $C_j \subset C_{j+1}$, $j=1,\dots,p-1$. The members of $\mathcal{K}^p(M)$ are called cornets (of length p on a set M).

If $C \in \mathcal{K}^p(M)$, then C_j, $j \in \{1,\dots,p\}$, denotes the jth coordinate of the cornet C, i.e. $C = \{C_j\}_{j=1}^p$.

If $V \in \mathcal{K}^{p+1}(M)$, we define $\mathcal{U}(V) = \{ C \in \mathcal{K}^p(M) \mid V_j \subset C_j \subset V_{j+1}, j=1,\dots,p \}$.

Now let $\{D_i\}_{i=1}^j$, $j \in \{1,\dots,p\}$ be a sequence of subsets of M. Let $C \in \mathcal{K}^p(M)$. We define $C - \{D_i\}_{i=1}^j$ to be the cornet $\tilde{C} \in \mathcal{K}^p(M)$ satisfying $\tilde{C}_t = C_t - \bigcup_{i=t}^j D_i$, $t=1,\dots,p$.

Let $V \in \mathcal{K}^{p+1}(M)$. We define $V \triangledown \{D_i\}_{i=1}^j = \{ C \in \mathcal{K}^p(M) \mid C \in \mathcal{U}(V - \{D_i\}_{i=1}^j)$ and $C_i \cap D_i = \emptyset$, $i=1,\dots,j \}$.

Remarks: The definition of $C - \{D_i\}$ is really correct. $V \triangledown \{D_i\}_{i=1}^j \subset \mathcal{U}(V - \{T_i\}_{i=1}^{j+1})$ where $T_i = D_i \cup D_{i-1}$, $i=2,\dots,j$, $T_{j+1} = D_j$, $T_1 = D_1$.

<u>Notation</u>: Let Q be a set, $F \in \mathcal{P}(Q)$, $C \in \mathcal{K}^{p+1}(M)$, $j \in \{1,\dots,p\}$. Let $r : \mathcal{P}(\mathcal{K}^p(M)) \longrightarrow \mathcal{P}(Q)$ be a mapping and let ξ be an infinite regular cardinal less than $|M|$. $A(p,j,F,C)$ denotes the following formula (where the sets X_i and Y_i in $A(p,j,F,C)$ are members of $[M]^{\leq \xi}$) $\forall Y_{j+1} \exists X_j \supset Y_{j+1} \forall Y_j \supset X_j \exists X_{j-1} \supset Y_j \forall Y_{j-1} \supset X_{j-1} \dots \exists X_1 \supset Y_2 \forall Y_1 \supset X_1 : r((C - \{Y_i\}_{i=1}^{j}) \nabla \{X_i\}_{i=1}^{j}) \subset F$.

<u>Basic lemma</u>: Let n be a positive integer. Let M be an uncountable set and let ξ be a regular infinite cardinal less than $|M|$. Let Q be a set. Let $r : \mathcal{P}(\mathcal{K}^n(M)) \longrightarrow \mathcal{P}(Q)$ be a mapping. If the following conditions (0), (1) are satisfied:

(0) for each pair X,Y : if $X \subset Y \subset \mathcal{K}^n(M)$, then $r(X) \subset r(Y)$, and

(1) there exist $j_0 \in \{1,\dots,n\}$ and $C \in \mathcal{K}^{n+1}(M)$ with $|C_1| = |M|$ such that the formula $A(n,j_0,F,C)$ is not valid for any $F \in [Q]^{\leq \xi}$ (i.e. $\exists j_0 \forall C \forall F$: non $A(n,j_0,F,C)$),

then there is $\tilde{C} \in \mathcal{K}^{n+1}(M)$ such that $|r(\mathcal{U}(\tilde{C}))| \geq \xi$.

In addition, we may suppose that $C_i = \tilde{C}_i$ for all $i > j_0$.

<u>Proof</u>: The Basic Lemma can be found in [P_3]. We omit the proof due to its length and complexity.

<u>Point-character of uniform box-product</u>

We are going to show that there is a very simple construction of an α-box product which yields uniform spaces of large point-character.

<u>Definition</u>: Given a uniform space $(X, \mathcal{V})$, an infinite cardinal α, and a non-empty index set I, we define a uniform α-box product $\prod_\alpha X^I = (X^I, \mathcal{V}_\alpha^I)$ as a uniform space whose underlying set is X^I and the basis of the uniformity $\mathcal{V}_\alpha^I$ is formed by all covers of the form:
$$\bigwedge_{s \in S} \pi_s^{-1}(\mathcal{P}) \quad \text{where } S \in [I]^{<\alpha} \text{ and } \mathcal{P} \in \mathcal{V}.$$

<u>Remarks</u>: 0) The uniform α-box product of a zerodimensional uniform space is 0-dimensional.

1) $\prod_\alpha + R^\alpha$ (where R denotes the uniform space of real numbers) in-

duces the usual uniformity on $\ell_\infty(\alpha)$, $(\ell_\infty(\alpha) \subsetneqq R^\alpha)$.

2) One can define the uniform α-box product in a more general setting: it is not necessary to suppose that all coordinate spaces are equal to each other. Even then the following theorem remains valid (the assumption of the following Theorem would then read that at least α coordinate spaces are not 0-dimensional).

Theorem: Let $(X, \mathcal{V})$ be a uniform space that is not 0-dimensional. Let α be an infinite cardinal. If $|I| \geq \alpha$ then $pc(\mathcal{T}_\alpha + X^I) > \xi$ for each regular cardinal $\xi < \alpha$.

Proof: The following lemmas are needed.

Definition: A finite sequence $\{M_i\}_{i=1}^n$ of sets is a chain of length n if: 1) $M_i \cap M_j \neq \emptyset$ iff $|i-j| \leq 1$,

2) $M_{i+1} - \bigcup_{t=1}^{i} M_t \neq \emptyset$ for $i=1,\ldots,n-1$.

Lemma 1: Let $(X, \mathcal{V})$ be a uniform space. The following conditions are equivalent: 1) $(X, \mathcal{V})$ is not 0-dimensional;

2) there is $\mathcal{O} \in \mathcal{V}$ such that for each $\mathcal{S} \in \mathcal{V}$, there is a chain $\{S_i\}_{i=1}^n$ of members of $\mathcal{S}$ such that $S_1 \cup S_n$ is not contained in any member of $\mathcal{O}$;

3) there is a cover $\mathcal{O} \in \mathcal{V}$ such that there is $\mathcal{P} \in \mathcal{V}$, $\mathcal{P} < \mathcal{O}$ such that for each $\mathcal{R} \in \mathcal{V}$, $\mathcal{R} < \mathcal{P}$ and each $\mathcal{S} \in \mathcal{V}$, $\mathcal{S} < \mathcal{R}$ there is a chain $\{S_i\}_{i=1}^n$ of elements of $\mathcal{S}$ such that $st(S_1, \mathcal{P}) \cap st(S_n, \mathcal{P}) = \emptyset$.

Proof of Lemma 1: $(3) \Rightarrow (1)$ is selfevident as $non(1) \Rightarrow non(3)$.

$(2) \Rightarrow (3)$. Take $\mathcal{P} \in \mathcal{V}$, $\mathcal{P} \overset{++}{<} \mathcal{O}$ where $\mathcal{O}$ is the cover guaranteed by (2).

$(1) \Rightarrow (2)$. We show: $non(2) \Rightarrow non(1)$. Suppose that for each $\mathcal{P} \in \mathcal{V}$ there is $\mathcal{L}_\mathcal{P} \in \mathcal{V}$ such that for each chain $\{S_i\}_{i=1}^n$ of elements of $\mathcal{L}_\mathcal{P}$, $S_1 \cup S_n$ is contained in some member of $\mathcal{P}$. Choose $\mathcal{S} \in \mathcal{V}$. We show that there is a uniform refinement of $\mathcal{S}$ of order 1. Choose a uniform cover $\mathcal{T}$ such that $\mathcal{T} \overset{+}{\leq} \mathcal{S}$. Consider $\mathcal{L}_{\mathcal{T}}$. We define a relation $\rho \subset X \times X$ by $(x_1, x_2) \in \rho$ iff there is a chain $\{S_i\}_{i=1}^{n}$ of elements of $\mathcal{L}_{\mathcal{T}}$ so that $x_1 \in S_1$, $x_2 \in S_n$. Evidently, ρ is reflexive. Its symmetry and transitivity is given by the following

<u>Lemma 2</u>: Let $\{T_i\}_{i=1}^{n}$ be a system of sets satisfying: $T_i \wedge T_{i+1} \neq \emptyset$, $i=1,\ldots,n-1$. Let $x \in T_1$, $y \in T_n$. Then there is a chain $\{S_j\}_{j=1}^{k}$ such that $\{S_j \mid j=1,\ldots,k\} \subset \{T_i \mid i=1,\ldots,n\}$ and $x \in S_1$, $y \in S_k$.

Hence ρ is an equivalence relation that induces a partition $\mathcal{D}$ of X. Evidently, $\mathcal{L}_{\mathcal{T}} < \mathcal{D}$, so $\mathcal{D} \in \mathcal{V}$ and it is easy to check that $\mathcal{D} < \mathcal{S}$ (use $\mathcal{T} \overset{+}{\leq} \mathcal{S}$). QED.

<u>Lemma 3</u>: Let $(X, \mathcal{V})$ be a uniform space. Let $\mathcal{P} \in \mathcal{V}$ be a cover of $\deg \mathcal{P} = \alpha$. There is $q \in \mathcal{V}$, $q < \mathcal{P}$ and $\mathcal{L} \in \mathcal{V}$ such that each member of $\mathcal{L}$ intersects less then α elements of q.

<u>Proof</u>: Apply the concept of a strict uniform shrinking ([I], Lemma VII. 3).

<u>Proof of Theorem</u>: We proceed by contradiction. Suppose that ($\odot$) : $pc(\Pi_\alpha + X^I) \leq \xi$ for some regular cardinal $\xi < \alpha$. Choose $M'' \subset I$, $|M''| = \alpha$. Let $\mathcal{O} \in \mathcal{V}$ and $\mathcal{P} \in \mathcal{V}$ be covers whose existence is given by Lemma 1 (3). Denote $\mathcal{X} = \bigwedge_{m \in M''} \Pi_m^{-1}(\mathcal{P})$, hence $\mathcal{X} \in \mathcal{V}^I_{\alpha^+}$. By $\odot$ and Lemma 3, there is a uniform cover $\mathcal{Y} = \bigwedge_{m \in M'} \Pi_m^{-1}(\mathcal{R})$, $M' \in [I]^{\leq \alpha}$, $\mathcal{R} \in \mathcal{V}$ such that $\mathcal{Y} < \mathcal{X}$ and there is $\mathcal{W}' \in \mathcal{V}^I_{\alpha^+}$ such that each $W \in \mathcal{W}'$ intersects less than ξ elements of $\mathcal{Y}$. Using Lemma 1 (3) and properties of a uniformity, we obtain a uniform cover $\mathcal{W} = \bigwedge_{m \in M} \Pi_m^{-1}(\mathcal{S})$ $\mathcal{W} \in \mathcal{V}^I_{\alpha^+}$, such that $\mathcal{W} < \mathcal{Y}$ and $\mathcal{W} \overset{++}{\leq} \mathcal{W}'$ and there is a chain $\{S_i\}_{i=1}^{n}$ of elements of $\mathcal{S}$ such that $st(S_1, \mathcal{P}) \cap st(S_n, \mathcal{P}) = \emptyset$ (it follows that the length of this chain is at least 4). Since $\mathcal{W} < \mathcal{Y}$, $M \supset M' \supset M''$, so $|M| = \alpha$. Choose $x_o \in X$. We define $Z = \{y \in X^I \mid ((m \in I-M) \Longrightarrow (y_m = x_o))$ and $((m \in M) \Longrightarrow (y_m \in \bigcup_{i=1}^{n} S_i))\}$.

Define $\varphi: Z \longrightarrow \mathcal{K}^{n-1}(M)$ by $\varphi(y) = C$ iff $C_j = \{m \in M | y_m \in \bigcup_{i=1}^{j} S_i\}$ $j=1,\ldots,n-1$.

Observation 1: φ is onto $\mathcal{K}^{n-1}(M)$.

Proof: Use the properties of chains.

Observation 2: Let $V \in \mathcal{K}^n(M)$. For $x,y \in \varphi^{-1}(\mathcal{U}(V))$ and $m \in M$, the following holds: if $x_m \in S_j$ and $y_m \in S_k$ them $|j-k| \leq 1$.

Proof: Suppose $j < k$, hence $j < n-1$. $V_{j+1} \supset \{\iota \in M | x_\iota \in \bigcup_{i=1}^{j} S_i\}$, so $m \in V_{j+1}$. But $V_{j+1} \subset \{\iota \in M | y_\iota \in \bigcup_{i=1}^{j+1} S_i\}$, hence $y_m \in S_{j+1}$.

Observation 3: For each cornet $V \in \mathcal{K}^n(M)$, the set $\varphi^{-1}(\mathcal{U}(V))$ intersects less than ξ elements of $\mathcal{Y}$.

Proof: Let $x \in \varphi^{-1}(\mathcal{U}(V))$. For each $m \in M$, choose $i(m) \in \{1,\ldots,n\}$ so that $x \in \bigcap_{m \in M} \pi_m^{-1}(S_{i(m)})$. Set $\bigcap_{m \in M} \pi_m^{-1}(S_{i(m)}) = W_x$. Evidently, $W_x \in \mathcal{W}$. By Observation 2, $\varphi^{-1}(\mathcal{U}(V)) \subset st(W_x, \mathcal{W})$. Now use $\mathcal{W}^{*\!*} \preceq \mathcal{W}'$ and the fact that each member of $\mathcal{W}'$ intersects less than ξ elements of $\mathcal{Y}$.

Observation 4: Let $V^1, V^2 \in \mathcal{K}^n(M)$. If $M' \cap (V_1^1 - V_n^2) \neq \emptyset$ (recall that $\mathcal{Y} = \bigwedge_{m \in M'} \pi_m^{-1}(\mathcal{R})$), then there is no $Y \in \mathcal{Y}$ satisfying:
$Y \cap \varphi^{-1}(\mathcal{U}(V^1)) \neq \emptyset \neq Y \cap \varphi^{-1}(\mathcal{U}(V^2))$.

Proof: Put $Y = \bigcap_{m \in M'} \pi_m^{-1}(R(m))$, $R(m) \in \mathcal{R}$ for each $m \in M'$. Let $x \in$ $\in Y \cap \varphi^{-1}(\mathcal{U}(V^1))$. Let $m_o \in (V_1^1 - V_n^2) \cap M'$. Then $x_{m_o} \in S_1$. Let $y \in Y \cap \varphi^{-1}(\mathcal{U}(V^2))$. Then $y_{m_o} \in S_n - \bigcup_{i=1}^{n-1} S_i \neq \emptyset$. So $R(m_o) \cap S_1 \neq \emptyset \neq$ $\neq R(m_o) \cap S_n - \bigcup_{i=1}^{n-1} S_i$, which contradicts Lemma 1 (3) and $\mathcal{R} < \mathcal{P}$.

Now define a mapping $r: \mathcal{P}(\mathcal{K}^{n-1}(M)) \longrightarrow \mathcal{P}(\mathcal{Y})$ by $r(\mathcal{D}) = \{Y \in \mathcal{Y} |$ there is $D \in \mathcal{D}$ such that $\varphi^{-1}(D) \cap Y \neq \emptyset\}$ for $\mathcal{D} \subset \mathcal{K}^{n-1}(M)$.

Observation 5: If $V \in \mathcal{K}^n(M)$ satisfies: $|V_1| = \alpha$, $V_n = M'$ then $A(n,n-1,F,V)$ does not hold for any $F \in [\mathcal{Y}]^{\leq \xi}$.

Proof: Perform an easy prolonged computation using Observation 4. Hence the assumptions of the Basic Lemma are satisfied, so there is $\tilde{V} \in \mathcal{K}^n(M)$ such that $|(r(\mathcal{U}(\tilde{V})))\geq \xi$, which is a contradiction.

Remarks: 1) The Theorem shows that $pc(l_\infty(\alpha)) > \xi$, where ξ is a regular cardinal less than α ; in particular, $pc\, l_\infty(\omega_1) > \omega_0$. The point character of $l_\infty(\omega_0)$ is an open problem, but we feel that it should soon be solved.

2) By similar methods we have partially solved a problem concerning the preservation of Cauchy filters by reflections in Unif (see [P_4], [P_5]). This problem is due to Z.Frolík and particular cases are mentioned in [I],[GI]). Our main result says thet if F is a reflection preserving Cauchy filters, then the spaces in $\{F(l_\infty(\alpha)) | \alpha \in Cn\}$ do not have bounded point-character.

Cardinal modifications

Definition: Let ω_α be a cardinal. We define a functor p^α : UNIF$\rightarrow$ $\rightarrow$UNIF by $p^\alpha(X,\mathcal{V}) = (X,p^\alpha\mathcal{V})$ where $p^\alpha\mathcal{V}$ consists of all covers $\mathcal{P}_0 \in \mathcal{V}$ such that there is a sequence $\{\mathcal{P}_n\}_{n=1}^{\infty} \subset \mathcal{V}$ with $|\mathcal{P}_n| < \omega_\alpha$ for n=1,2,... and $\mathcal{P}_n \overset{+}{<} \mathcal{P}_{n-1}$, n=1,2,... .

Remark: p^α is a reflection that preserves underlying sets and topology. Such reflections are called modifications.

Definition: Let $(X,\mathcal{V})$ be a uniform space. Let ω_α be a cardinal. We define $b^\alpha\mathcal{V} = \{\mathcal{P} \in \mathcal{V} \mid |\mathcal{P}| < \omega_\alpha\}$.

Remark: Clearly, $(X,b^\alpha\mathcal{V})$ is a quasiuniformity in the sense of [I]. The difficulties are connected with star-refinements. If $(X,b^\alpha\mathcal{V})$ is a uniformity, then $(X,b^\alpha\mathcal{V}) = p^\alpha(X,\mathcal{V})$.

It is well-known that $(X,b^0\mathcal{V})$ and $(X,b^1\mathcal{V})$ always form uniformities. A more general theorem, proved in [V] and [K], says: if $pc(X,\mathcal{V}) \leq \omega_0$, then $(X,b^\alpha\mathcal{V})$ is a uniformity for any cardinal ω_α . On the other hand, A.Kucia proved under [GCH] : Let $(X,\mathcal{V})$ be a uni form space. Then $(X, b^{\alpha}\mathcal{V})$ forms a uniformity for any cardinal ω_α. Both these theorems are corollaries of the Folklore Lemma introduced below. Since there are uniform spaces with large point-character it is possible that the equality $p^{\alpha}(X,\mathcal{V}) = (X, b^{\alpha}\mathcal{V})$ depends on set--theoretical assumptions for $\alpha \geq 2$. It is really the case.

Notation: The symbol $S^{+}(M)$ denotes the positive unit sphere in $\ell_\infty(M)$, i.e. the subspace of $\ell_\infty(M)$ on the set $\{f \in \ell_\infty(M) \mid \|f\| = 1$ and $f(m) \geq 0$ for each $m \in M\}$.

Notation: For each $f \in S^{+}(M)$ and a non-negative integer k, define $C^{(f,k)} \in \mathcal{K}^{2^k}(M)$ by $C_i^{(f,k)} = f^{-1}(\,]\!]\frac{2^k - i}{2^k}, 1]\!]\,)$, $i = 1, \ldots, 2^k$. For $V \in \mathcal{K}^{2^k+1}(M)$, put $\mathcal{W}(V) = \{f \in S^{+}(M) \mid C^{(f,k)} \in \mathcal{U}(V)\}$.

Definition: For a non-negative integer k, we define $\mathcal{P}_k = \{\mathcal{W}(V) \mid V \in \mathcal{K}^{2^k+1}(M)\}$.

Proposition: $\{\mathcal{P}_k\}_{k=1}^{\infty}$ forms a basis for the norm uniformity on $S^{+}(M)$.

Proof (see $[P_5]$).

Notation: The mapping which assigns $C^{(f,k)} \in \mathcal{K}^{2^k}(M)$ to each $f \in S^{+}(M)$ will be denoted by φ_k.

Let ω_α be an uncountable cardinal. $ZB(\omega_\alpha)$ denotes the following assertion: There is $\mathcal{A} \subset [\omega_\alpha]^{\omega_\alpha}$, $|\mathcal{A}| > \omega_\alpha$, an infinite regular cardinal $\xi < \mathrm{cf}\,\omega_\alpha$, and a cardinal $K \leq |\mathcal{A}|$ such that $|\bigcap \mathcal{A}'| < \xi$ for each $\mathcal{A}' \in [\mathcal{A}]^{\geq K}$.

Remark: [GCH] implies that $ZB(\omega_\alpha)$ is false.

Theorem: If $ZB(\omega_\alpha)$ holds, then there is a uniform cover $\mathcal{P}$ of $S^{+}(\omega_\alpha)$ such that $|\mathcal{P}| = \omega_\alpha$ and $\mathcal{Q} \overset{*}{<} \mathcal{P}$ implies $|\mathcal{Q}| > \omega_\alpha$ (i.e. $p^{\alpha}S^{+}(\omega_\alpha) \neq b^{\alpha}S^{+}(\omega_\alpha)$).

The following lemma is needed.

Folklore Lemma $[P_6]$: Let $(X, \mathcal{V})$ be a uniform space. Let K be an infinite cardinal. Let $\mathcal{P} = \{P_a\}_{a \in A} \in \mathcal{V}$, $|\mathcal{P}| < K$. Let $\mathcal{R} = \{R_b\}_{b \in B}$ be a uniform star-refinement of $\mathcal{P}$. For $x \in X$ put

$S(x) = \{a \mid a \in A$ and $st(x, \mathcal{R}) \subset P_a\}$. For $Y \subset X$, put $I(Y) = \{a \mid a \in A$ and $Y \subset P_a\}$. A mapping $t : X \longrightarrow A$ satisfying $t(x) \in S(x)$ for each $x \in X$ will be called a choice mapping.

Assertion of Folklore Lemma: There is $\mathcal{Q} \in \mathcal{V}$, $\mathcal{Q} = \{Q_a\}_{a \in A}$, such that $\mathcal{Q} \overset{+}{<} \mathcal{P}$ and $\mathcal{R} < \mathcal{Q}$ iff the following condition (P) is satisfied:

(P) There is a choice mapping $t : X \longrightarrow A$ and a partition $\{B_a\}_{a \in A}$ of the index set B such that: $t(\bigcup_{b \in B_a} R_b) = \bigcup_{b \in B_a} t(R_b) \subset$
$\subset \bigcup_{b \in B_a} I(R_b) = I(\bigcup_{b \in B_a} R_b)$.

Proof of Theorem: For $a \in \omega_\alpha$, put $\tilde{a} = \{f \in S^+(\omega_\alpha) \mid a \in \mathrm{coz}\, f\}$. Define $\mathcal{P} = \{\tilde{a}\}_{a \in \omega_\alpha}$. Clearly, $\mathcal{P}$ is a uniform cover of $S^+(\omega_\alpha)$. Suppose that there is a uniform cover $\mathcal{Q} = \{Q_\iota\}_{\iota \in \omega_\alpha}$ such that $\mathcal{Q} \overset{+}{<} \mathcal{P}$. By the proposition, there is a $k \geq 2$ such that $\mathcal{P}_k < \mathcal{Q}$. By the Folklore Lemma, there is a choice mapping $t : S^+(\omega_\alpha) \longrightarrow \omega_\alpha$ such that $st(f, \mathcal{P}_k) \subset \widetilde{t(f)}$ for each $f \in S^+(\omega_\alpha)$ and a partition $\{B_a\}_{a \in \omega_\alpha}$ such that the following is satisfied for each $a \in \omega_\alpha$:

(*) $\bigcup_{V \in B_a} t(\mathcal{U}(V)) \subset \bigcap_{V \in B_a} I(\mathcal{U}(V))$ (I was defined in the Folklore Lemma).

Clearly, for each $f \in S^+(\omega_\alpha)$, there is $g \in st(f, \mathcal{P}_k)$ such that $\mathrm{coz}\, g = f^{-1}(]2^{-k}, 1]) = C^{(f,k)}_{2^k - 1}$; hence: (since $st(f, \mathcal{P}_k) \subset \widetilde{t(f)}$):

(**) $t(f) \in C^{(f,k)}_{2^k - 1}$.

Define $r : \mathcal{P}(\mathcal{K}^{2^k}(\omega_\alpha)) \longrightarrow \mathcal{P}(\omega_\alpha)$ by $r(\mathcal{D}) = t(\{f \in S^+(\omega_\alpha) \mid C^{(f,k)} \in \mathcal{D}\})$. We obtain from (**) and $C^{(f,k)}_{2^k - 1} \subset V_{2^k}$ for each f: (***) $r(\mathcal{W}(V)) \subset V_{2^k}$ for each $V \in \mathcal{K}^{2^k+1}(\omega_\alpha)$. Let $\mathcal{A}$ be a collection of sets whose existence is given by $ZB(\omega_\alpha)$. For $L \in \mathcal{A}$, take a cornet $V^L \in \mathcal{K}^{2^k+1}(\omega_\alpha)$ such that $V^L_{2^k+1} = \omega_\alpha$, $V^L_{2^k} = L$ and $|V_1| = \omega_\alpha$.

It is clear from (**) that $A(2^k, 2^k - 1, F, V^L)$ does not hold for any

$F \in [\omega_\alpha]^{\leq \xi}$, so we can use the Basic Lemma: there is $\tilde{V}^L \in \mathcal{K}^{2^k+1}(\omega_\alpha)$ such that $|\tilde{V}^L_{2^k+1}| = \omega_\alpha$, $\tilde{V}^L_{2^k} = L$ and $|r(\mathcal{W}(V^L))| \geq \xi$ (we have used correspondence given by φ_k) .

Since there is $f \in \mathcal{W}(\tilde{V}^L)$ such that coz $f = L$ we also have $I(\mathcal{W}(\tilde{V}^L)) = L$.

Now consider the restriction of $\{B_a\}_{a \in A}$ to $\{\tilde{V}^L | L \in \mathcal{a}\}$. Since $|\mathcal{a}| > \omega_\alpha$, there is a_o such that $|\{L \in \mathcal{a} | \tilde{V}^L \in B_{a_o}\}| \geq |\mathcal{a}| \geq K$.

Denote $\mathcal{a}' = \{L \in \mathcal{a} | \tilde{V}^L \in B_{a_o}\}$. By (*) we have:

$$\bigcup_{L \in \mathcal{a}'} t(\mathcal{W}(\tilde{V}^L)) = \bigcup_{L \in \mathcal{a}'} r(\mathcal{W}(\tilde{V}^L)) \subset \bigcap_{L \in \mathcal{a}'} I(\mathcal{W}(\tilde{V}^L)) = \bigcap_{L \in \mathcal{a}'} L = \bigcap \mathcal{a}' .$$

According to $ZB(\omega_\alpha)$, $|\bigcap \mathcal{a}'| < \xi$, although $|r(\mathcal{W}(\tilde{V}^L))| \geq \xi$ for each $L \in \mathcal{a}$, which is a contradiction.

<u>Comment</u>: We have mentioned that [GCH] implies the negation of $ZB(\omega_\alpha)$. One could doubt whether $ZB(\omega_\alpha)$ is consistent with ZFC. Fortunately, [B] removes these unpleasant questions.

<u>Notation</u>: $AB(K, \lambda, \mu, \nu)$ denotes the assertion: there is $F \subset [K]^\mu$ such that $|F| = \lambda$ and $|X \cap Y| < \nu$ if $X, Y \in F$ and $X \neq Y$.

<u>Theorem Baumgartner</u> [B]: It is consistent with ZFC to suppose that $AB(K, \lambda, K, \nu)$ holds, where $\nu \leq K \leq \lambda$ and ν is regular.

<u>Remark</u>: Clearly, $AB(\omega_\alpha, \omega_\alpha^+, \omega_\alpha, \xi)$, where ξ is a regular cardinal less than cf ω_α , implies $ZB(\omega_\alpha)$.

So we see that the assertion: "For each uniform space $(X, \mathcal{V})$ and each $\alpha \geq 2$, $p^\alpha(X, \mathcal{V}) = (X, b^\alpha \mathcal{V})$" is consistent with and independent of ZFC.

References

[B] Baumgartner J.E.: Almost disjoint sets, the dense-set problem and the partition calculus, Annals of Math.Log., 1976 (4).

[GI] Ginsburg S., Isbell J.R.: Some operators on uniform spaces, Trans.A.M.S., 93 (1959).

[I] Isbell J.R.: Uniform spaces, Mathematical Surveys (12), A.M.S., 1964.

[K] Kulpa W.: On uniform universal spaces, Fund.Math.LXIX (1970).

[P_1] Pelant J.: Remark on locally fine spaces, Comment.Math.Univ. Car., 16 (1975).

[P_2] Pelant J.: General hedgehogs in general topology, Seminar Uniform Spaces direct by Z.Frolík, Matematický ústav ČSAV, Praha, 1976.

[P_3] Pelant J.: Cardinal reflections and point-character of uniformities, Seminar Uniform Spaces 1973-74, directed by Z.Frolík Matematický ústav ČSAV, Praha, 1975.

[P_4] Pelant J.: Reflections not preserving completeness, ibid.

[P_5] Pelant J.: Point character of uniformities and completeness, Seminar Uniform Spaces 1975-76 directed by Z.Frolík, Matematický ústav ČSAV, Praha, 1976.

[P_6] Pelant J.: One folkloristic lemma on cardinal reflections, Seminar Uniform Spaces 1973-74 directed by Z.Frolík, Matematický ústav ČSAV, Praha, 1975.

[PPV] Pelant J., Preiss D., Vilímovský J.: On local uniformities,to appear in Gen.Top. and its Appl.

[RR] Reynolds G.D., Rice M.D.: Completeness and covering properties of uniform spaces, to appear.

[S] Stone A.H.: Universal spaces for some metrizable uniformities,

Quart.J.Math., 11 (1960).

[V] Vidossich G.: A note on cardinal reflections in the category of uniform spaces, Proc.A.M.S., 23 (1969).

NONDISCRETE MATHEMATICAL INDUCTION

Vlastimil PTÁK
Czechoslovak Academy of Sciences
Institute of Mathematics
Žitná 25, 115 67 Praha 1, Czechoslovakia

The lecture is divided into the following sections:

1. Motivation
2. Statement of the induction theorem
3. Relation to classical theorems
4. Principles of application
5. An illustration: the factorization theorem
6. New results
7. Connections with numerical analysis

1. Motivation

This lecture presents a report about a series of investigations whose aim it is to set up an abstract model for iterative existence proofs and constructions in analysis and numerical analysis.

I intend to show that a model which describes a large class of iteration processes may be based on a certain modification of the closed graph theorem.

Let us start with the following observation. In existence proofs in mathematical analysis and in numerical analysis we often devise iterative procedures in order to construct an element which lies in a certain set or satisfies a given relation. At each stage of the iterative process we are dealing with elements which satisfy the desired relation only approximately, the degree of approximation becoming better at each step.

To describe the abstract model which we shall investigate later, consider the problem of constructing a point x which belongs to a given set W. We start by replacing the given set W by a family $W(r)$ of sets depending on a small positive parameter r ; the inclusion $z \in W(r)$ means - roughly speaking - that the inclusion $z \in W$ is satisfied only approximately, the approximation being measured by the number r. All the $W(r)$ are supposed to be subsets of a complete metric space (E,d).

In what follows we intend to show that, under suitable hypotheses concerning the relation between the sets $W(.)$ and the metric of the space a simple theorem may be proved which gives the construction of an iterative process converging to a point $x \in W$. The theorem, the so-called induction theorem, is closely related to the closed graph theorem in functional analysis; it could be described as a quantitative strengthening of the closed graph theorem. Indeed, the closed graph theorem can be viewed, in a certain sense, as a limit case of the induction theorem, for an infinitely fast rate of convergence. The proof of the induction theorem is an exercise; the interest of the result lies exclusively in its formulation, which makes it possible to unify a number of theorems in one simple abstract result.

2. Metric spaces and the Induction Theorem

Definition. Let T be an interval of the form $T = \{t; 0 < t < t_o\}$ for some positive t_o. A rate of convergence or a small function on T is a function ω defined on T with the following properties

1° ω maps T into itself

2° for each $t \in T$ the series $t + \omega(t) + \omega^{(2)}(t) + \ldots$ is convergent.

We use the abbreviation $\omega^{(n)}$ for the n-th iterate of the function ω, so that $\omega^{(2)}(t) = \omega(\omega(t))$ and so on. The sum of the above series will be denoted by σ. The function σ satisfies the following functional equation

$$\sigma(t) - t = \sigma(\omega(t)) ;$$

one of the consequences of this fact is the possibility of recovering ω if σ is given. Indeed, we have

$$\omega(t) = \sigma^{-1}(\sigma(t) - t)$$

(with the exception of pathological cases).

Given a metric space (E,d) with distance function d, a point $x \in E$ and a positive number r, we denote by $U(x,r)$ the open spherical neighbourhood of x with radius r, $U(x,r) = \{y \in E ; d(y,x) < r\}$. Similarly, if $M \subset E$, we denote by $U(M,r)$ the set of all $y \in E$ for which $d(y,M) < r$. If we are given, for each sufficiently small positive r, a set $A(r) \subset E$, we define the limit $A(0)$ of the family $A(.)$ as follows

$$A(0) = \bigcap_{s>0} \left(\bigcup_{r \leq s} A(r)\right)^- .$$

Now we may state the Induction Theorem.

Theorem. Let (E,d) be a complete metric space, let T be an interval $\{t ; 0 < t < t_o\}$ and ω a rate of convergence on T. For each $t \in T$ let $Z(t)$ be a subset of E ; denote by $Z(0)$ the limit of the family $Z(.)$. Suppose that

$$Z(t) \subset U(Z(\omega(t)), t)$$

for each $t \in T$. Then

$$Z(t) \subset U(Z(0, \sigma(t))$$

for each $t \in T$.

Proof. An exercise.

3. Relation to classical theorems

It will be interesting to compare the induction theorem with some classical results. It is almost immediate that the Banach fixed point principle is a simple consequence.

Let E be a complete metric space and f a mapping of E into itself such that

$$d(f(x_1), f(x_2)) \leq \alpha d(x_1,x_2) \quad \text{for all} \quad x_1,x_2 \in E$$

where α is a fixed number, $0 < \alpha < 1$. Then there exists an $x \in E$ such that $x = f(x)$.

Proof. For each $t > 0$ set

$$Z(t) = \{x ; d(x,f(x)) < t\} ;$$

it follows that $Z(0) = \{x ; x=f(x)\}$. It will be sufficient to show that $Z(t) \subset U(Z(\alpha t),t)$. If $x \in Z(t)$, set $x' = f(x)$ so that $d(x,x') < t$. Let us show that $x' \in Z(\alpha t)$. This, however, is immedia since

$$d(x',f(x')) = d(f(x),f(x')) \leq \alpha d(x,x') = \alpha d(x,f(x)) < \alpha t .$$

The induction theorem applies with $\omega(r) = \alpha r$.

The connection with the closed graph theorem is somewhat less obvious. Roughly speaking, the closed graph theorem consists in the following implication: a mapping which is uniformly almost open is already open. Now the induction theorem can be described as a quantitative refinement of the closed graph theorem. To see that, let us recall the notion of a uniformly almost open mapping [4].

A mapping f from a uniform space E into a uniform space V is said to be uniformly almost open, if, for each entourage U in E, there exists an entourage V in F such that, for each x, we have $f(U(x))^- \supset V(f(x))$. This means, that points of $V(f(x))$ may be arbitr rily well approximated by points from $f(U(x))$. The conclusion is that

- under appropriate hypotheses about the spaces and the mapping - that, for a slightly larger $U' \supset U$ we already have the inclusion $f(U'(x)) \supset$ $\supset V(f(x))$ for all x .

It turns out that the same conclusion can be obtained under a weaker assumption. The approximability of V by the elements of $f(U)$ need not be arbitrarily good. It suffices if we are able to approximate to a finite distance, provided the error of the approximation is small as compared with the size of the entourages. Smallness is measured by a small function; the conclusion also gives an information how much larger U' has to be in order to have the inclusion $f(U') \supset V$. For details, see the author's remark [4].

4. Principles of application

Now we should explain why the method has been given the name of nondiscrete mathematical induction. We shall see that the application of the method consists in reducing the given problem to a system of functional inequalities for several indeterminate functions one of which is to be a rate of convergence; this explains the word nondiscrete. The connection with the classical method of mathematical induction is obvious - we investigate the possibility of passing from a point x which approximates the point to be constructed with an error not exceeding r to another point x' close to x for which the approximation is considerably better.

Suppose we are given an approximation of order r , in other words, a point $x \in W(r)$ and are allowed to move from x to a distance not greater than r . Can we find, within $U(x,r)$, an approximation of a much better order r'? A suitable way of giving this a precise meaning is to impose the condition $r' = \omega(r)$ where ω is a small function. The condition that for each $x \in W(r)$ there exists a point $x' \in U(x,r) \cap$ $\cap W(r')$ with $r' = \omega(r)$ may also be expressed in the form $W(r) \subset$ $\subset U(W(\omega(r)), r)$ so that $W(r)$ satisfies the hypotheses of the induction theorem. We have thus

$$W(t) \subset U(W(0), \sigma(t))$$

for sufficiently small t . Hence we shall be able to assert that $W(0)$ is nonvoid provided at least one $W(r)$ is nonvoid.

This corresponds to the first step of an ordinary induction proof; here, as in the discrete case, we have to make sure that the process begins somewhere. There is another point which should be stressed, the heuristic value of the method.

The main advantage consists in the fact that the iterative construc ion is taken care of by the general theorem so that the application consists in the verification of the hypotheses, the main question being: how much can a given approximation be improved within a given neighbourhood. By separating the hard analysis portion from the construction of the sequence, this method not only yields considerable simplifications of proofs but also evidences more clearly the substance of the problem. Instead of defining an approximation process first and then investigati the degree of approximation at the n-th step the method we propose coul be described as exactly the opposite: we begin by looking at the sets $W(r)$ where the degree of approximation is at least r, then choose a suitable rate of convergence; the induction principle gives the constru ion of an iterative sequence corresponding to that rate of convergence automatically.

In this manner, we are using the relation between the improvement of the approximation and the distance we have to go in order to attain it in the most advantageous manner. There are examples to show that a given system of functional inequalities may be consistent with differen rates of convergence. The conclusion obtained from the Induction Theore may differ according to the choice of these; however, there seems to be (at least in the concrete problems investigated thus far - in particula in the case of the Newton process, which we shall discuss later) a natu rate of convergence which yields the best possible result - in the sens that the estimates are sharp within the class of problems under conside ation.

Now let us give all this a more precise formulation.

Let (E,d) be a complete metric space and f a nonnegative continuous function on E. We are looking for a point x for which $f(x) = 0$.

1st observation. Let us assume that, for each x taken from some set $M \subset E$ and each positive $r < r_0$ we can prove an estimate of the form

$$\inf \{f(x'); x' \in U(x,r)\} < h(f(x), r)$$

where h is a suitable function of two variables. Suppose there exists a positive function φ tending to zero with r and a rate of converge ω such that

$$h(\varphi(r), r) \leq \varphi(\omega(r))$$

Set $W(r) = \{x \in M , f(x) \leq \varphi(r)\}$; then $W(r) \subset U(W(\omega(r)), r)$.

2nd observation. The functional equation connecting ω and σ may be used to obtain information about the distance of the solution from any point u_0 given in advance. Indeed, let u_0 be a fixed poin in E . Given a point $x \in E$ and two positive numbers d and r such that

$$d(x,u_o) \leq d - \sigma(r) ,$$

then, for $x' \in U(x,r)$, we have

$$d(x',u_o) \leq d(x,u_o) + d(x',x) \leq d - \sigma(r) + r = d - \sigma(\omega(r)) .$$

It follows that the family

$$Z(r) = \{x \in M ; f(x) \leq \varphi(r) , d(x,u_o) \leq d - \sigma(r)$$

satisfies $Z(r) \subset U(Z(\omega(r)), r)$. It follows that $Z(0)$ will be nonvoid if at least one $Z(r_o)$ is nonvoid since

$$Z(r_o) \subset U(Z(0), \sigma(r_o)) .$$

Summing up: if $h(\varphi(r),r) \leq \varphi(\omega(r))$ and if there exists an $r_o > 0$ and an $x_o \in M$ such that

$$f(x_o) \leq \varphi(r_o) \qquad d(x_o,u_o) \leq d - \sigma(r_o)$$

then there exists an $x_\infty \in M^-$ with the following properties

$$f(x_\infty) = 0$$
$$d(x_\infty,x_o) \leq \sigma(r_o)$$
$$d(x_\infty,u_o) \leq d .$$

We have seen that the first step of the induction method consists in finding a function $h(m,r)$ with the following property: given x with $f(x) \leq m$, there exists, within distance less than r , an x' with $f(x') \leq h(m,r)$. In most cases the estimate for $f(x')$ will not depend on the value of $f(x)$ alone but will require some further characteristics as well; one might think of derivatives or some other additional information.

Suppose, for simplicity, that there is only one such additional characteristic, i.e. that the estimate for $f(x')$ depends also on the value of another positive function f_1 at x so that $\inf \{f(x') , x' \in U(x,r)\} \leq h(f(x), f_1(x), r)$. Consider the case where the estimate h is an increasing function of the second argument. Since we shall need, in the following step of the induction an estimate for $f_1(x')$, we shall need, in fact, a pair of positive functions h and h_1 such that, for each x and r , there exists an $x' \in U(x,r)$ for which

$$f(x') \leq h(f(x), f_1(x),r)$$
$$f_1(x') \geq h_1(f(x),f_1(x),r).$$

In this case h_1 will have to be decreasing in the first argument and increasing in the second argument. It will then be desirable to find a pair of functions φ, φ_1 and a rate of convergence ω such that

$$h(\varphi(r), \varphi_1(r), r) \leq \varphi(\omega(r))$$
$$h_1(\varphi(r), \varphi_1(r), r) \geq \varphi_1(\omega(r)) .$$

Let $W(r) = \{x \in E ; f(x) \leq \varphi(r), f_1(x) \geq \varphi_1(r)\}$; then $W(r) \subset U(W(\omega(r)), r)$.

Let us pass now to examples which illustrate the general principles sketched above.

5. The factorization theorem

The method of nondiscrete mathematical induction has been applied thus far to obtain improvements of selection theorems, transitivity theorems in the theory of C^*-algebras, factorization theorems in Banach algebras and existence theorems in the theory of partial differential equations. The first three are described in the author's paper [3]. The ideas contained there have also been applied successfully by the author's collaborators [10], [11].

Among the many examples which demonstrate the advantages of the method the Rudin-Cohen factorization theorem seems to be the most suitable one; in spite of the fact that the result itself is not new the simplification of the proof is considerable.

If M is a unital Banach algebra, we denote by $G(M)$ the set of its invertible elements. Let A be a Banach algebra without unit and denote by B its unitization. The multiplicative functional on B whi[ch] has A as its kernel will be denoted by f. Let F be a Banach space which is an A-module. We say that the pair (A,F) possesses an approxi[-] mate unit of norm β if, for each $a \in A$, $y \in F$ and $\varepsilon > 0$ there exists an $e \in A$ such that

$$|e| \leq \beta \quad , \quad |ea - a| < \varepsilon \quad , \quad |ey - y| < \varepsilon$$

Theorem. Let A be a Banach algebra without a unit and let F b[e] a Banach space which is an A-module. Suppose that (A,F) possesses an approximate unit of norm β. Then, for each $y \in F$ and each $\varepsilon > 0$, there exists an $a \in A$ and a $z \in F$ such that

$$az = y \ , \quad |a| \leq \beta \ , \quad z \in (Ay)^- \ , \quad |z - y| \leq \varepsilon \ .$$

Proof. First of all, it is easy to see that the existence of a bounded approximate unit implies $(Ay)^- = (By)^-$ for any $y \in F$, B being the unitization of A.

Consider the space $A \times (By)^-$ equipped with the norm

$$\|p\| = \frac{1}{1 - \omega} \max \left\{ \frac{1}{\beta}|a| \ , \ \frac{1}{\varepsilon}|z| \right\}$$

if $p = [a,z]$; $0 < \omega < 1$ is a constant to be chosen later. For each invertible $b \in B$ let $p(b)$ be the pair $p(b) = [a,z]$, $a = b^{-1} - f(b^{-1}$[)], $z = by$. For each $r > 0$, set

$$W(r) = \{p(b);\ b \in G(B),\ |f(b^{-1})| \leq r,\ \|p(b) - p(1)\| \leq \frac{1}{1-\omega}(1-r)\} \ .$$

In particular, $[0,y] = p(1) \in W(1)$. Also, observe that $[a,z] \in W(r)$ implies

$$az = (b^{-1} - f(b^{-1}))by = y - f(b^{-1})z$$

so that $[a,z] \in W(0)$ implies $az = y$.

We intend to show that there exists an ω such that

$$W(r) \subset U(W(\omega r), r) \quad \text{for each} \quad r > 0 .$$

Having proved that, it will follow from the Induction Principle that $W(1) \subset U(W(0), \frac{1}{1-\omega})$; this means that there exists a pair $p = [a,z] \in W(0)$ with $\|p-p(1)\| \le \frac{1}{1-\omega}$, in other words $az = y$, $|a| \le \beta$, $|z-y| \le \varepsilon$. Given $p(b) \in W(r)$ we intend to show that the pair $p(b')$ corresponding to a slightly perturbed $b' = bc$ will satisfy $p(b') \in W(\omega r)) \cap U(p(b), r)$. For this, clearly it suffices to construct c in such a manner that

(1) $\quad |a' - a| < (1 - \omega)\beta r$

(2) $\quad f(c^{-1}) = \omega$

(3) $\quad |(b' - b)y| \le (1 - \omega)\varepsilon r$.

We shall show that it is possible to satisfy these three conditions by constructing a c for which

(4) $\quad b'^{-1} - b^{-1}$ is a multiple of $e - 1$ for a suitable e .

Since $f(e-1) = -1$, such a choice of c - if possible - has the following consequences:

$$a' - a = b'^{-1} - b^{-1} - f(b'^{-1} - b^{-1}) = (1 - \omega)f(b^{-1})e$$
$$b' - b = -b'(b'^{-1} - b^{-1})b = -b'(1 - \omega)f(b^{-1})(e-1)b$$
$$b'(1 + (1 - \omega)f(b^{-1})(e-1)b) = b ;$$

for shortness, write $w = (1 - \omega)f(b^{-1})(e-1)b$.

It follows that a suitable choice of c will be $c = (1+w)^{-1}$ provided $|w| < 1$.

We shall need an estimate for w independent of e . Let τ be such that $(1 - \omega)(\beta + 1) = 1 - 2\tau$. Since

$$w = (1 - \omega)f(b)^{-1}(e-1)(b-f(b)) + (1 - \omega)(e-1)$$

we have

$$|w| \le (1 - \omega)|f(b)^{-1}| \; |(e-1)(b-f(b))| + (1 - \omega)(\beta + 1)$$

and $e \in A$, $|e| \le \beta$ may be chosen so as to have

(5) $\quad (1 - \omega)|f(b)^{-1}| \; |(e-1)(b-f(b))| + (1 - \omega)(\beta + 1) \le 1 - \tau$

whence $|w| \le 1 - \tau$ and $|c| \le \frac{1}{\tau}$.

Now suppose that e satisfies condition (5) and at the same time

$$|b| \tfrac{1}{\tau}(1 - \omega)|f(b)^{-1}| \; |(e-1)by| < (1 - \omega)\varepsilon r$$

then

$$|z' - z| < (1 - \omega)\varepsilon r .$$

The proof is complete.

6. New results

The classical notion of the order of convergence or rate of convergence which reputedly goes back to the last century is defined as follows Given an iterative process which yields a sequence x_n of elements of a complete metric space (E,d) converging to an element $x \in E$ we say that the convergence is of order p if there exists a constant α such that

$$d(x_{n+1},x) \leq \alpha(d(x_n,x))^p \ .$$

Clearly it is immaterial whether we require this for all n or only asymptotically. Let us point out two difficulties which seem to arise if this point of view is adopted.

1° If $p > 1$ then the above inequality contains a certain amount of information about the process; the information, however, is more of a qualitative nature since it relates quantities which we are not able to measure at any finite stage of the process. The obvious meaning of th above inequality seems to consist rather in the fact that, at each stage of the process, the following step of the iteration yields a significant improvement of the estimate.

2° Theoretical considerations enable us, in many cases, to establi an inequality of the above type for certain constants α and p ; however, usually this is only possible if we assume n to be larger than a certain bound. We might want, however, to stop the process before this bound is reached - in this case the inequality cannot be used. Of course it is possible to extend the validity of the estimate to all n by maki α sufficiently large - this may invalidate its practical applicability for the initial steps.

It seems therefore reasonable to look for another method of estimat ing the convergence of iterative processes, one which would satisfy the following requirements.

1° It should relate quantities which may be measured or estimated during the actual process.

2° It should describe accurately in particular the initial stage of the process, not only its asymptotic behaviour since, after all, we are interested in keeping the number of steps necessary to obtain a goo estimate as low as possible.

It is obvious that we cannot expect to have an adequate descriptio of both the beginning and the tail end of the process by any formula as simple as the one we discussed above. In our opinion, a description which fits the whole process, not only an asymptotic one, is only possible by means of suitable functions, not just numbers.

It seems natural to expect that better results may be obtained by looking for small functions ω which relate two consecutive increments of the process by an inequality of the following type

$$d(x_{n+1},x_n) \leq \omega(d(x_n,x_{n-1})) .$$

By allowing a larger class of functions than just those of the type $t \to \alpha t^p$ we have a better chance of getting a closer fit of the estimates even at the beginning of the process.

At the same time this approach measures the rate of convergence at finite stages of the process using only data available at that particular stage of the process, in fact, instead of comparing the two unknown quantities $d(x_n,x)$ and $d(x_{n+1},x)$ it is based on the relation between $d(x_n,x_{n-1})$ and $d(x_{n+1},x_n)$.

Suppose we have a sequence of inequalities

$$d(x_{n+k},x_{n+k-1}) \leq \omega^{(k-1)}(d(x_{n+1},x_n))$$

for $k=1,2,\ldots$ (where $\omega^{(j)}$ stands for the j-th iteration of the function ω) and that the series $\sum_0^\infty \omega^{(j)}(d(x_{n+1},x_n))$ is convergent. Such a sequence of inequalities may be deduced from the above inequality if ω is an increasing function. Then the sequence $x_n,x_{n+1},\ldots$ is a fundamental sequence and, the space (E,d) being complete, converges to a limit x for which

$$d(x_n,x) \leq d(x_{n+1},x_n) + d(x_{n+2},x_{n+1}) + \ldots \leq \sum_0^\infty \omega^{(j)}(d(x_{n+1},x_n)) .$$

As an example, let us mention the rate of convergence of Newton's process recently established by the author. There we have

$$\omega(t) = \frac{t^2}{2(t^2 + d)^{1/2}}$$

where d is a positive constant depending on the data of the problem. A closer inspection of this formula shows that, for every small t , the function assumes approximately the form $\frac{t^2}{2d^{1/2}}$ whereas, for large t , the summand t^2 predominates in the denominator so that the function is approximately linear, $\frac{1}{2}t$.

Since ω relates the consecutive steps of Newton's process by the inequality $d(x_{n+1},x_n) \leq \omega(d(x_n,x_{n-1}))$ this shows first that, asymptotically - in other words for small $d(x_n,x_{n-1})$ - the next increment is approximately $\frac{1}{2d^{1/2}}(d(x_n,x_{n-1}))^2$. This phenomenon is usually described by saying that the convergence is quadratic.

However, in the initial stages of the process $d(x_n, x_{n-1})$ is still large so that ω is almost linear. Since it may be shown that the estimates for Newton's process based on ω are sharp at each step, it follow that accurate estimates valid for the whole process - including the initial steps - cannot be based on any simple quadratic monomial.

The precise formulation is as follows.

Let E and F be two Banach spaces, let $x_0 \in E$ and $U = \{x ; |x-x_0| \leq m\}$. Let f be a mapping of U into F twice Fréche differentiable for each $x \in U$. Suppose the following conditions are satisfied:

1° there exists a constant $k > 0$ such that $|f''(x)| \leq k$ for all $x \in U$,

2° $f'(x_0)$ is invertible and $d(f'(x_0)) = d_0 > 0$

3° $|f'(x_0)^{-1} f(x_0)| \leq r_0$.

If $d_0 \geq 2kr_0$ and if

$$m \geq \frac{d_0}{k}\left(1 - \left(1 - \frac{2kr_0}{d_0}\right)^{1/2}\right)$$

then the Newton process starting at x_0 is meaningful and converges to a point x such that $f(x) = 0$. The function

$$\omega(r) = \frac{r^2}{2(r^2+d)^{1/2}}$$

where $d = \left(\frac{d_0}{k}\right)\left(1 - \frac{2kr_0}{d_0}\right)$, is a rate of convergence and yields the following estimates

$$|x_{n+1} - x_n| \leq \omega^{(n)}(r_0)$$

$$|x - x_0| \leq \frac{d_0}{k}\left(1 - \left(1 - \frac{2kr_0}{d_0}\right)^{1/2}\right).$$

These estimates are sharp in the following sense: for each triple k, d_0, r_0 of positive numbers satisfying the inequality $d_0 \geq 2kr_0$ there exists a mapping f for which these estimates are attained.

The proof is given in [6]. The corresponding σ function is comput in [6] and the finite sums σ_n in [7].

Let us conclude this section by mentioning another example a detail discussion of which may be found in [5].

If γ and β are positive numbers such that $\gamma > 4\beta$ then

$$\omega(t) = t\,\frac{\gamma + t - ((\gamma + t)^2 - 4\beta)^{1/2}}{\gamma - t + ((\gamma + t)^2 - 4\beta)^{1/2}}$$

is a rate of convergence on the whole positive axis. It has been used in [5] to obtain a result on the spectrum of an almost decomposable operato The corresponding σ-function is computed in [5] and the finite sums σ_n

in [8].

7. Connections with numerical analysis

Let us turn now to the problem of comparing this new method of measuring convergence with the classical notion described at the beginning. The new method is based on comparing consecutive terms in the sequence

$$d(x_n, x_{n+1})$$

while the classical one compares consecutive terms in the sequence

$$d(x_n, x) \ .$$

It is thus natural to ask whether estimates using consecutive distances $d(x_n, x_{n-1})$ imply similar estimates for the distances $d(x_n, x)$. More precisely, if $e_{n,n+1}$ stands for an estimate of $d(x_{n+1}, x_n)$ and e_n for an estimate of $d(x_n, x)$ we can ask whether estimates of the form $e_{n+1,n+2} \leq \omega(e_{n,n+1})$ imply estimates of the classical type $e_{n+1} \leq \leq \omega(e_n)$. We intend to show that this is indeed so at least in the case where ω is convex.

To see that, suppose we have a sequence x_n for which the estimate

$$d(x_{n+1}, x_n) \leq \omega(d(x_n, x_{n-1}))$$

holds. Hence

$$d(x_n, x) \leq d(x_{n+1}, x_n) + d(x_{n+2}, x_{n+1}) + \dots \leq$$

$$\leq \sum_0^{\infty} \omega^{(k)}(d(x_{n+1}, x_n)) = \sigma(d(x_{n+1}, x_n)) \ .$$

Here we have used the fact that ω is nondecreasing; this is a simple consequence of the convexity of ω.

Similarly, $d(x_{n+1}, x) \leq \sigma(\omega(d(x_{n+1}, x_n)))$; it follows that the estimates

$$e_{n+p,n+p+1} = \omega^{(p)}(d(x_{n+1}, x_n)) \qquad p=0,1,2,\dots$$

and

$$e_n = \sigma(e_{n,n+1})$$

satisfy the inequalities $e_{n+1} \leq \sigma(\omega(e_{n,n+1}))$.

To obtain the desirable estimate $e_{n+1} \leq \omega(e_n)$ it would be sufficient to have the inequality $\sigma \circ \omega \leq \omega \circ \sigma$ since this yields the following estimates

$$e_{n+1} \leq \sigma(\omega(e_{n,n+1})) \leq \omega(\sigma(e_{n,n+1})) = \omega(e_n) \ .$$

This heuristic reasoning should be sufficient to explain the importance of the inequality $\sigma \circ \omega \leq \omega \circ \sigma$.

It turns out that such an inequality may be proved in the case of convex rates of convergence ω. The following proposition holds [12].

Suppose ω is a rate of convergence on the interval T. If ω is convex then

$$\omega \circ \sigma \geq \sigma \circ \omega$$

on the interval $T \cap \sigma^{-1}(T)$.

It follows that, in this case, the two ways of estimating convergen discussed above are equivalent.

Detailed proofs and a discussion of the basic principles of the non-discrete induction method may be found in the Gatlinburg Lecture [5].

List of references

[1] V. PTÁK, Some metric aspects of the open mapping theorem, Math. Ann. 163 (1966), 95-104

[2] V. PTÁK, Deux théorèmes de factorisation, Comptes Rendus Acad.Sci. Paris 278 (1974),1091-1094

[3] V.PTÁK, A theorem of the closed graph type, Manuscripta Math. 13 (1974), 109-130

[4] V. PTÁK, A quantitative refinement of the closed graph theorem, Czechoslovak Math.J. 99 (1974), 503-506

[5] V. PTÁK, Nondiscrete mathematical induction and iterative existenc proofs, Linear Algebra and Appl. 13 (1976), 223-238

[6] V. PTÁK, The rate of convergence of Newton's process, Numer.Math. 25 (1976), 279-285

[7] V. PTÁK, Concerning the rate of convergence of Newton's process, Comment.Math.Univ.Carolinae 16 (1975), 699-705

[8] V. PTÁK, A rate of convergence, Abh.Math.Sem.Univ. Hamburg (in print)

[9] V. PTÁK, A modification of Newton's method, Časopis pěst.mat. 101 (1976), 188-194

[10] H. PETZELTOVÁ and P. VRBOVÁ, A remark on small divisors problems, Revue Roumaine Math. (in print)

[11] J. ZEMÁNEK, A remark on transitivity of operator algebras, Časopis pěst.mat. 100 (1975), 176-178

[12] V. PTÁK, What should be a rate of convergence, RAIRO, Analyse numérique (in print)

COMPACT C-SPACES AND S-SPACES

by

M. Rajagopalan

ABSTRACT.

We introduce a set theoretic axiom $♣_\infty$ which is weaker than $♣$ as well as axiom F. Using (CH) and $♣_\infty$ we prove the existence of a locally compact, T_2, locally countable, first countable, hereditarily separable, sequentially compact non-compact space X. The one point compactification X^* of X is a compact, T_2, C-space (meaning X^* is of countable tightness) which is not sequential. We also construct a compact, T_2, C-space Y which is not sequential using only the continuum hypothesis (CH). This solves some well known problems on S-spaces and also on compact C-spaces under least set theoretic axioms.

INTRODUCTION.

Some areas of current interest in topology are cardinal functions and the role of set theoretic axioms. Much literature has grown around these topics. (See [1,2,4,5,7,8,9,10,11]). Set theoretic axioms are used mainly to construct examples like S-spaces. An S-space is a hereditarily separable completely regular space which is not Lindelöf. Spaces which come close to being an S-space are C-spaces in the sense of Mrowka and Moore [6] or, in the language of cardinal functions, spaces X whose tightness t(X) is countable. If Y is a space, the tightness t(Y) of Y is the least among the cardinals λ with the property that if $A \subset Y$ and $x_o \in \bar{A}$ then there is a subset $B \subset A$ of cardinality λ so that $x_o \in \bar{B}$. A C-space is a space Y whose tightness $t(Y) \leq \aleph_o$. The sequential spaces are C-spaces. A space S is called sequential if given $A \subset X$ we can get $\bar{A}$ by iterating the operation of taking limits of convergent sequences beginning from A. We give a more elaborate definition of sequential spaces below. The following problem has been raised several times by A.V. Arhangelskii and also by V. Kannan [5] and Ponomorov [11]. The problem is:

"IS A COMPACT, T_2, C-SPACE SEQUENTIAL?"

The first ones to raise a related problem are S. Mrowka and C.C. Moore [6] who asked whether a Hausdorff C-space is sequential. An example of a Hausdorff C-space which is not sequential was given by Franklin and Rajagopalan and they raised the problem whether a regular, C-space must be sequential. (See [3]).

The above problem of Kannan and Arhangelskii on compact, T_2, C-spaces can be answered in the negative by assuming strong set theoretic axioms. Thus using continuum hypothesis (CH) and the axiom $\clubsuit$ Ostazewski [7] constructed a locally compact, T_2, sequentially compact, first countable, locally countable, hereditarily separable noncompact space X. Such a space was also constructed by Fedorchuk in [2] using axiom F which is stronger than both (CH) and $\clubsuit$.

So the hard question is "what are the least set of axioms which guarantee the existence of such S-spaces as the ones constructed by Ostazewski or which guarantee the existence of compact, T_2, C-spaces which are not sequential?"

In this paper we introduce an axiom $\clubsuit_\infty$ which is weaker than $\clubsuit$. We show that (CH) and $\clubsuit_\infty$ together imply the existence of an S-space such as the one got by Ostazewski. We also show that assuming (CH) alone; there is a compact, T_2, C-space which is not sequential.

NOTATIONS.

We consider only Hausdorff spaces. We assume ZFC which is Zermelo-Frankael set theoretic aximos with axiom of choice. If we use axioms beyond ZFC in set theory in any of our lemmas or theorems we mention only those axioms in the hypothesis of those lemmas or theorems. We follow [12] for basic notions in topology. N is the set of integers $> o$ with discrete topology and βN is its Stone-Čec compactification. Ω is the first uncountable ordinal. If A,B are sets then A/B is the set difference A - B. (CH) denotes the continuum hypothesis. We follow [10] for statements of the axioms (CH), $\diamondsuit$, $\clubsuit$, (MA) and $\circledast$. If X is a topological space and π a partition of X then X/π denotes the quotient space of X given by π. The axiom (F) is stated in Fedorchuk [2].

DEFINITION 1.

Let X be a topological space. Let $A \subset X$. A is called sequentially open if no sequence lying in X/A converges to an element of A. X is called sequential if and only if every sequentially open subset A of X is open. X is called a C-space if given $A \subset X$ and an element $x_o \in \overline{A}$ there is a countable subset $B \subset A$ so that $x_o \in \overline{B}$.

AXIOM $\clubsuit_n$.

Let n be a given integer $> o$. The axiom $\clubsuit_n$ is the following: For every limit ordinal α in $[1, \Omega)$ there are n sets $A_{\alpha 1}$,

$A_{\alpha 2}, \ldots A_{\alpha n}$ so that the following hold:

(a) $A_{\alpha i} \subset [1,\alpha)$ for $i=1,2,..,n$.

(b) $A_{\alpha i}$ is cofinal with $[1,\alpha)$ for all $i=1,2,\ldots,n$.

(c) Given an uncountable subset $M \subset [1,\Omega)$ there exists $\alpha < \Omega$ and $i \in \{1,2,\ldots,n\}$ so that $M \supset A_{\alpha i}$.

AXIOM $\clubsuit_F$.

This is the following statement. Given a limit ordinal $\alpha \in [1,\Omega)$ there is an integer n_α and sets $A_{\alpha 1}, A_{\alpha 2}, \ldots, A_{\alpha n_\alpha}$ satisfying the following:

(i) $A_{\alpha i} \subset [1,\alpha)$ and is cofinal with $[1,\alpha)$ for all α in $[1,\Omega)$ and $i=1,2,..,n_\alpha$.

(ii) Given an uncountable subset $B \subset [1,\Omega)$ there is an $\alpha \in [1,\Omega)$ and an 'i' so that $1 \leq i \leq n_\alpha$ such that $A_{\alpha i} \subset B$.

AXIOM $\clubsuit_\infty$.

This is the following statement. Given a limit ordinal α in $[1,\Omega)$ there exist sets $A_{\alpha 1}, A_{\alpha 2}, \ldots, A_{\alpha n}, \ldots$ so that the following hold:

(I) $A_{\alpha n} \subset [1,\alpha)$ and is cofinal with α for all $n \in N$.

(II) Given an uncountable subset $B \subset [1,\Omega)$ there is an ordinal $\alpha \in [1,\Omega)$ and an integer $n \in N$ so that $A_{\alpha n} \subset B$.

We notice that the axiom $\clubsuit$ of Ostazewski is our $\clubsuit_1$. Clearly $\clubsuit$ implies $\clubsuit_n$ and $\clubsuit_n$ implies $\clubsuit_F$ for all $n \in N$. Moreover $\clubsuit_F$ is easily seen to imply $\clubsuit_\infty$. It is natural to ask whether any of these implications is reversible. But we do not go into it here. The axiom (F) implies $\clubsuit$.

We proceed to prove the following two theorems here:

THEOREM I.

(CH) + $\clubsuit_\infty$ imply that there exists a locally compact, T_2, first countable, hereditarily separable, sequentially compact, locally countable non-compact space S. The one point compactification S^* of S is a compact, T_2, C-space which is not sequential.

THEOREM II.

(CH) alone implies the existence of a compact, T_2, C-space which is not sequential.

We begin to prove Theorem I. We use V-process. The V-process is described in [10]. We begin with the following lemma:

DEFINITION 2.

Let $\underline{A}$ be a countable collection of closed sets of βN and Y an open dense subset of βN such that $\bigcup_{X \varepsilon \underline{A}} X \subset Y$. $\underline{A}$ is called a discrete collection in Y if for every subcollection $\underline{B}$ of $\underline{A}$ we have that $\bigcup_{X \varepsilon \underline{B}} X$ is closed in Y.

DEFINITION 3.

If $\alpha \varepsilon [1, \Omega]$ then λ_α denotes the αth limit ordinal in $[1, \Omega]$. In other words $\lambda_\alpha = \omega^\alpha$ for all $\alpha \varepsilon [1, \Omega]$ where ω is the least limit ordinal in $[1, \Omega]$.

Now we will follow the V-process method of [10] with a slight alteration. For this we will define a closed non-empty subset $B\gamma$ of βN for each ordinal γ in $[1, \Omega]$ so that $B_\gamma \cap B_\delta = \phi$ if $\gamma \neq \delta$ and $\gamma, \delta < \Omega$ and $\bigcup_{\gamma < \alpha} B_\gamma$ is a dense open subset of βN for all limit ordinals $\alpha < \Omega$. Then the collection $\{B_\gamma | \gamma \varepsilon [1, \Omega]\}$ will give a partition π of $Y_{\Omega-} = \bigcup_{\gamma < \Omega} B_\gamma$ and $Y_{\Omega-} / \pi$ will be the required locally compact space of Theorem I.

DEFINITION 4.

For every $n \varepsilon N$ we put $B_n = \{n\}$. Put $\underline{A_1}$ to be the collection of all infinite subsets of $\{B_n | n \varepsilon N\}$. $\underline{A_1}$ is well ordered as $A_{11}, A_{12}, \ldots, A_{1\delta}, \ldots$ using (CH) where $\delta \varepsilon [1, \Omega]$. $Y_1 = N = \bigcup_{n<\omega} B_n$ and $\pi_1 = \{\{n\} | n \varepsilon N\}$ by definition.

LEMMA 5.

Let Y be a dense open subset of βN and π a partition of Y by compact sets of βN. Let $\{n\} \varepsilon \pi$ for $n \varepsilon N$. Let the following be satisfied:

(a) Y/π is locally compact and T_2.

(b) Given $A \varepsilon \pi$ there is a compact, open subset V of βN so that V is a countable union of members of π and $A \subset V$.

Let $\underline{A_n} = \{P_{n1}, P_{n2}, \ldots, P_{nr}, \ldots\}$ be a countably infinite collection of members of π so that $\bigcup_{n=1}^{\infty} \underline{A_n} = \{P_{ij} | i,j \varepsilon N\}$ is a discrete collection in Y (see Definition 2). Then there are non-empty compact subsets $C_1, C_2, \ldots, C_n, \ldots$ of βN so that the following hold:

(i) $Y \cup C_n$ is open in βN for all $n \varepsilon N$.

(ii) $C_n \neq \phi$ and $C_n \cap C_m = \phi$ for all $n, m \varepsilon N$ so that $n \neq m$.

(iii) Given $n \varepsilon N$ there is a compact open set V_n of βN so that $C_n \subset V_n$ and V_n/C_n can be expressed as a countable union of members of π.

(iv) Given $n, b \varepsilon N$ and a compact open subset W of βN containing C_n we have that $W \supset P_{ki}$ for some $i \varepsilon N$.

PROOF.

Let $N_1, N_2, \ldots, N_k, \ldots$ be a pairwise disjoint collection of infinite subsets of N so that $\bigcup_{k=1}^{\infty} N_k = N$. Given $n, r \varepsilon N$ find a compact open subset V_{nr} of βN so that $P_{nr} \subset V_{nr} \subset \beta N$ and V_{nr} is a union of countably many members of π and $V_{nr} \cap V_{ms} = \phi$ if either $n \neq m$ or $r \neq s$ for all $n, m, r, s \varepsilon N$. Such a family V_{nr} is easily seen to exist by the hypothesis ⓐ and the discreteness of $\underline{\bigcup_{n=1}^{\infty} A_n}$. Let $W_i = \bigcup_{n=1}^{\infty} (\bigcup_{r \varepsilon N_i} V_{nr})$ for all $i \varepsilon N$. Let $C_n = \overline{W}_n - W_n$ for all $n \varepsilon N$. Then, this family $C_1, C_2, \ldots, C_n, \ldots$ is the required family of sets.

DEFINITION 6.

Let $A \subset \beta N$. The growth A^* of A is defined as $\overline{A}/A$.

LEMMA 7.

Let $Y \subset \beta N$ be a dense open set and π a partition of Y by compact sets so that Y/π is locally compact, T_2 and countable. Assume further that given a member $A \varepsilon \pi$ there is a compact open set W of βN so that $A \subset W \subset Y$ and W is a union of members of π. Let $A_1, A_2, \ldots, A_n, \ldots$ be a sequence of distinct members of π so that the growth A^* of the set $A = \bigcup_{n=1}^{\infty} A_n$ is non-empty and disjoint with Y. Then there is a dense open set M of βN with the following properties:

(a) M/Y is a non-empty compact open set.

(b) If $\pi_o = \pi \cup \{M/Y\}$ then π_o is a partition of M so that M/π_o is a countable, locally compact, T_2, space.

(c) There is a compact open set W of βN so that $(M/Y) \subset W \subset M$ and $W \cap Y$ is a union of members of π.

(d) $A^* \cap M \neq \phi$.

Proof.

This is proved in [10].

LEMMA 8.

Let Y and π be as in the hypothesis of Lemma 7. Let (A_n) be a sequence of families of members of π as in the Lemma 6. Let (F_n) be a sequence of infinite collections of members of π so that the growth A_n^* of the set $A_n = \bigcup_{X \in F_n} X$ is non-empty and disjoint with Y for all $n \in N$. Then there is a sequence $D_1, D_2, \ldots, D_n, \ldots$ of non-empty compact sets of βN so that the following hold:

(a) $Y \cup D_n$ is open in βN for all $n \in N$.

(b) There exists a compact open set V_n of βN so that $D_n \subset V_n \subset (Y \cup D_n)$ and $V_n \cap Y$ is a union of members of π for all $n \in N$.

(c) $D_n \cap D_m = \phi$ for all $n,m \in N$ and $n \neq m$.

(d) Given $n,k \in N$ and a compact open set W of βN containing D_n there is a member A of A_k so that $A \subset W$.

(e) If $M = Y \cup (\bigcup_{n=1}^{\infty} D_n)$ and $\pi_o = \pi \cup \{D_n | n \in N\}$ then π_o is a partition of M and M/π_o is a countable, locally compact, T_2 space.

(f) $M \cap A_n^* \neq \phi$ for all $n \in N$ where A_n is defined above in this Lemma.

Proof.

First of all get compact sets $C_1, C_2, \ldots, C_n, \ldots$ as in the conclusion of Lemma 5. Put $Y_1 = Y \cup (\bigcup_{n=1}^{\infty} {}_n)$ and $\pi_1 = \pi \cup \{C_1, C_2, \ldots, C_n, \ldots\}$. Then (Y_1, π_1) satisfy the hypothesis of Lemma 7. If $A_n^* \cap Y_1 \neq \phi$ for all $n \in N$ then take $D_n = C_n$ for all $n \in N$. If not let n_1 be the first integer so that $A_{n_1}^* \cap Y_1 = \phi$. Apply Lemma 7 and an open set M as in the conclusion of that lemma with A_{n_1} replacing A and (Y_1, π_1) replacing (Y, π) in that lemma. Then there is a compact set F of $\beta N - N$ so that $F \neq \phi$ and $F \cap Y_1 = \phi$ and there is a compact open set W of βN so that $F \subset W \subset (F \cup Y_1)$ and $W \cap Y_1$ is a union of members of π_1 and $A_{n_1}^* \cap W \neq \phi$. A look at the proof of Lemma 5 shows that W can be further chosen so that $W \cap C_n = \phi$ for all $n \in N$. So choose an open compact subset W as above, Put $D_1 = C_1 \cup (W/Y_1)$. We define D_n in general by induction. Assume that

n is a given integer > 1 and that we have defined $D_1, \ldots, D_{n-1}$ in such a way that $A_i^* \cap (Y_1 \cup \bigcup_{i=1}^{n-1} D_i) \neq \phi$ for $i=1,2,\ldots,n-1$. Put

$Y_2 = Y_1 \cup (\bigcup_{i=1}^{n-1} D_i)$ and $\pi_2 = \pi \cup \{D_1, \ldots, D_{n-1}, C_n, C_{n+1}, \ldots\}$. If $A_i^* \cap Y_2 \neq \phi$ for all $i \in N$ then put $D_i = C_i$ if $i \geq n$. If not there is a least integer k so that $A_k^* \cap Y_2 = \phi$. Then we get a W_o as above with the condition that $W_o \cap D_i = \phi$ for $i=1,2,$
$W_o \cap C_i = \phi$ for $i \geq n$ and $W_o \cap A_k^* \neq \phi$ and $W_o \cap Y$ is a union of members of π. Put $D_n = W_o/Y_2$. Thus proceeding by induction we get $D_1, D_2, \ldots, D_n, \ldots$ as required.

LEMMA 9.

Let $\clubsuit_\infty$ be satisfied. For every countable limit ordinal α let $A_{\alpha 1}, A_{\alpha 2}, \ldots, A_{\alpha n}, \ldots$ be as in the statement of $\clubsuit_\infty$. Then, given a limit ordinal $\alpha \in [1, \Omega)$ there is a countable family F_α of subsets of $[1, \alpha)$ with the following properties:

(a) If $A \in F_\alpha$ then A is cofinal in $[1, \alpha)$.

(b) Given a limit ordinal α and $A, B \in F_\alpha$ we have $A \cap B = \phi$ unless $A = B$.

(c) If α is a limit ordinal and F_α is infinite and $F_\alpha = \{A_{\alpha 1}, A_{\alpha 2}, \ldots, A_{\alpha n}, \ldots\}$ and g_n is the least element in $A_{\alpha n}$ for $n \in N$ then we have $g_1 < g_2 < \ldots < g_n < \ldots$ and the sequence (g_n) is cofinal with α.

(d) If α is a limit ordinal and $A \in F_\alpha$ then A is discrete and closed in $[1, \alpha)$ in the usual order topology of $[1, \alpha)$.

(e) Given an uncountable subset $B \subset [1, \Omega)$ there is a countable limit ordinal α and a set $A \in F_\alpha$ so that $A \subset B$.

(f) F_α is infinite for all limit ordinals α in $[1, \Omega)$.

Proof.

The proof of this conbinatorial Lemma is long and is postponed to appear in another paper.

Hereafter, we use (CH) and $\clubsuit_\infty$ and F_α will be as in Lemma 9.

CONSTRUCTION 10. (V-PROCESS).

We put $Y_1, \pi_1, \underline{A_1}, \underline{A_{1\delta}}, B_1, B_2, \ldots, B_n, \ldots$ as in definition 4 for all $n \in N$ and $\delta \in [1, \Omega)$. Recall that given $\alpha \in [1, \Omega); \lambda_\alpha$ denotes the αth limit ordinal in $[1, \Omega)$. Assume that given an ordinal α in $[1, \Omega)$ such that $\alpha > 1$ we have defined $Y_\gamma, \pi_\gamma, \underline{A_\gamma}, \underline{A_{\gamma\delta}}$, for all $\gamma < \alpha$ and $\delta \in [1, \Omega)$ and B_γ for all $\gamma < \lambda_{\alpha*}$ so as to satisfy the following; where $\alpha* = \alpha$ if α is a limit ordinal and $\alpha*$ is predecessor of α otherwise:

(i) Y_γ is a dense open set of βN and π_γ is a partition of Y_γ by compact sets for all $\gamma < \alpha$.

(ii) $1 \leq \gamma < \delta < \alpha$ implies that $Y_\gamma \subset Y_\delta$ and $\pi_\gamma \subset \pi_\delta$.

(iii) If $\gamma \in [1, \alpha)$ and $A \in \underline{F}_{\lambda_\gamma}$ and $A = (\delta_1, \delta_2, \ldots, \delta_n, \ldots)$ then given τ such that $\lambda_\gamma < \tau < \lambda_\alpha$ and a compact open set W of βN so that $B_\tau \subset W$ we have that $W \supset B_{\delta_n}$ for some $n \in N$.

(iv) Y_γ / π_γ is a countable, locally compact, T_2 space for all γ in $[1, \alpha)$.

(v) $\pi_\gamma = \{B_\delta | \delta \in [1, \lambda_\gamma)\}$ for all $\gamma \in [1, \alpha)$.

(vi) Given $\gamma \in [1, \alpha)$ and $\delta < \lambda_\gamma$ there is a compact open set W of βN so that $B_\delta \subset W \subset Y_\gamma$ and W is a union of members of π_γ.

(vii) Given $\gamma \in [1, \alpha)$ we have that $\underline{A_\gamma}$ is the collection of all infinite families of members $C_1, C_2, \ldots, C_n, \ldots$ of π_γ so that the growth $C*$ of $C = \bigcup_{n=1}^{\infty} C_n$ is non-empty and has empty intersection with Y_γ.

(viii) Given $\gamma \in [1, \alpha)$ we have that $\underline{A_{\gamma 1}}, \underline{A_{\gamma 2}}, \ldots, \underline{A_{\gamma\delta}}, \ldots$ is a well ordering of $\underline{A_\gamma}$ by $[1, \Omega)$.

(ix) Given $\gamma \in [1, \gamma)$ and $\delta \in [1, \gamma)$ and $\beta \in [1, \omega^\gamma)$ we have that $Y_\gamma \cap A^*_{\delta\beta} \neq \phi$ where $A^*_{\delta\beta}$ is the growth of the set $A_{\delta\beta} = \bigcup_{X \in \underline{A_{\delta\beta}}} X$.

Then we define $Y_\alpha, \pi_\alpha, \underline{A_\alpha}, \underline{A_{\alpha\delta}}, B_\gamma$ as follows for $\delta \in [1, \Omega)$ and $\gamma \in [\lambda_\alpha, \lambda_{\alpha+1})$. Consider $F_{\underline{\delta\alpha}}$ and write its members as $\underline{A_1}, \underline{A_2}, \ldots, \underline{A_n}, \ldots$ Put $Y_{\alpha-} = \bigcup_{\gamma<\alpha} Y_\gamma$ and $\pi_{\alpha-} = \bigcup_{\gamma<\alpha} \pi_\gamma$. Let $\underline{C}$ denote the set of all $\underline{A_{\delta\beta}}$ so that $1 \leq \delta < \alpha$ and $1 \leq \beta < \omega^\alpha$ so that growth $A^* \cap Y_{\alpha-} = \phi$

where $A = \bigcup_{X \varepsilon A_{\delta\beta}} X$. Now using $Y_{\alpha-}, \pi_{\alpha-}, (A_n)$, and C in Lemma 8 at the appropriate places get $D_1, D_2, \ldots, D_n, \ldots$ as in the conclusion of that lemma so that (a) - (f) of that lemma are satisfied.
Put $B_{\lambda_\alpha} = D_1$ and $B_{\lambda_\alpha + n} = D_{n+1}$ for all $n \varepsilon N$.

Put $Y_\alpha = \bigcup_{\gamma<\alpha} Y_\gamma \cup \bigcup_{n=1}^{\infty} D_n$ and $\pi_\alpha = \{B_\gamma | \gamma \varepsilon [1, \lambda_{\alpha+1})\}$.

Put A_α to be the set of all infinite families $\{B_{\gamma_1}, B_{\gamma_2}, \ldots, B_{\gamma_n}, \ldots\}$ where $\gamma_n \varepsilon [1, \lambda_{\alpha+1})$ for all $n \varepsilon N$ and such that the growth $A^* \neq \phi$ and $A^* \cap Y_\alpha = \phi$ where $A = \bigcup_{n=1}^{\infty} B_{\gamma_n}$. Let $A_{\alpha 1}, A_{\alpha 2}, \ldots, A_{\alpha\delta}, \ldots$ be a well ordering of A_α by $[1, \Omega)$. Finally we put $Y_\Omega = \bigcup_{\alpha<\Omega} Y_\alpha$ and $\pi_{\Omega-} = \bigcup_{\alpha<\Omega} Y_\alpha$ and $\pi_{\Omega-} = \bigcup_{\alpha<\Omega} \pi_\alpha$ and $X_o = Y_{\Omega-}/\pi_{\Omega-}$.

THEOREM 11.

X_o is a locally compact, T_2, sequentially compact, first countable, locally countable non-compact space. Further X_o is hereditarily separable.

Proof.

The proof of the properties of X_o except that of hereditary separability is exactly like that of Theorem 1.8 and 1.9 in [10]. Now we come to the hereditary separability of X_o. Let F be an uncountable subset of X_o. Then there exists a unique uncountable subset $B \subset [1, \Omega)$ so that $\phi(B_\gamma) \varepsilon F$ if and only if $\gamma \varepsilon B$ where ϕ: $Y_{\Omega-} \longrightarrow X_o$ is the natural map. Then there is an $\alpha \varepsilon [1, \Omega)$ and a member $A \varepsilon F_{\lambda_\alpha}$ so that $A \subset B$. Let $Z_o = \{\phi(x_\gamma) | \gamma \varepsilon A\}$. Then $\overline{Z}_o/B$ is at most countable. Hence there is a countable dense subset in B. Thus the Theorem.

THEOREM 12.

Assuming (CH) + $\clubsuit_\infty$ there is a compact, T_2, C-space which is not sequential.

Proof.

The one point compactification X_o^* of X_o in Theorem 11 is easily seen to be such an example.

REMARK.

The proof of Theorem II given in the beginning of this paper is long and cannot be accommodated in this hours talk. So it will appear elsewhere.

ACKNOWLEDGEMENT.

The author thanks sincerely the National Science Foundation of U.S.A. for having given him a travel grant and thus enable him to attend the topology symposium in Prague and present this talk there.

REFERENCE

(1) A.V. Arhangelskii, "On cardinal invariants, General Topology and its relation to Modern Algebra and Analysis III", Proceedings of third Prague Topology Symposium (1971), 37-46.

(2) V.V. Fedorchuk, "On the cardinality of hereditarily separable compact Hausdorff spaces", Soviet. Math. Dokl 16 (1975), 651-655.

(3) S.P. Franklin and M. Rajagopalan, "Some examples in topology", Trans. Amer. Math. Soc. 155 (1971), 305-314.

(4) I. Juhasz, "Cardinal Functions in Topology", Mathematical Centre Tracts, Amsterdam, 1971.

(5) V. Kannan, "Studies in Topology", Thesis, Madurai University, Madurai (INDIA), 1970.

(6) R.C. Moore and G.S. Mrowka, "Topologies determined by countable objects", Notices of Amer. Math. Soc. 11 (1964), 554.

(7) A. Ostazewski, "On countably compact perfectly normal spaces", J. Lond. Math. Soc., (to appear).

(8) T.C. Pryzymusinski, "A Lindelöf space X such that X^2 is normal, but not paracompact", Fund. Math. 78 (1973) 291-296.

(9) M. Rajagopalan, "Scattered spaces III", to appear in J. Ind. Math. Soc.

(10) M. Rajagopalan, "Some outstanding problems in topology and the V-process", Categorical Topology, Mannheim (1975), Lecture notes in mathematics, Vol. 540, Springer-Verlag, Berlin, (1976) 500-517.

(11) M.E. Rudin, "Lecture notes in set theoretic topology", CBMS Lecture series, American Mathematical Society, Providence (R.I.) 1975.

(12) R. Vaidyanathaswamy, "Set theoretic topology", Chelsea, New York (1960).

A NARROW VIEW OF SET THEORETIC TOPOLOGY

M.E. RUDIN
Madison

Topology today is many different subjects. Even leaving aside algebraic and differential topology, general topology is hardly one topic.

For example, Starbird [1] recently proved the following: Suppose that f is a piecewise linear homeomorphism of a 3-simplex onto itself which is fixed on the boundary. Then there is a triangulation T of σ and a continuous family $h_t: \sigma \to \sigma$ ($t \in [0,1]$) of linear-with-respect-to- T homeomorphisms such that h_0 is the identity, h_1 is f, and h_t is the identity on the boundary of σ for all t. Although one is interested in producing continuum many homeomorphisms, this is obviously not a set theoretic problem. Starbird's proof involves a combination of finite combinatories and purely geometric pushing and pulling. Geometric topology is a difficult branch of general topology with highly developed techniques and broad applications throughout mathematics.

The geometric techniques are the ones which have proved most effective in handling problems in infinite dimensional topology(by which I mean the study of Hilbert space manifolds). The space of all closed subsets of the closed unit interval may sound set theoretic but Schori and West [2] proved that this space is the Hilbert cube and their solution as well as those solving related problems is very geometric.

Another area in general topology which is not set theoretic is one which I call "continua theory." An example of a fundamental unsolved problem in this area is: does every compact connected set in the plane which does separate the plane have the fixed point property. Without knowing the answer, one still speculates that set theoretic considerations will not play a major role in its solution. In all of the above areas the spaces in question are separable and metric, usually manifolds with a linear or geometric structure as well. The desired constructions are achieved by finite or countable processes and the interplay between cardinals is not a factor.

However, problems involving compactifications, Baire or Borel sets, metrizability, paracompactness, normality, and more generally comparison problems between closely related topological properties are almost all really problems in set theory. The most obvious such problems involve the comparison of cardinal functions on a topological space; but many more subtle problems also have set theoretic translations.

This area blossoms at the moment due to the recognition of the set theoretic nature of the problems, the availability of an effective collection of set theoretic tools, and the increased acquaintance of topologists with these tools. Almost every day someone tells me a new result answering a question posed in the literature. The area is almost impossible to survey because too much is going on; I find it extremely difficult to distinguish which results are most important and will be lasting. I tried two years ago to survey [3] primarily those results gotten by Wisconsin visitors. A list of over 100 problems was given; perhaps 2/3 of these have now been at least partially solved. Of course, some papers have particularly wide influence. Ostaszewski's construction [4] has been frequently used. Juhász's book [5] is basic. Šapirovskii's unification [6] of the proofs of so many theorems is often used to simplify and clarify; Hodel [7] and Pol [15] have nice write ups of this. Efimov's broad paper [8] on extremal disconnectedness contains a wealth of material. The recent paper by Chaber [9] proving that countably compact spaces with a G_δ diagonal are metric is brief but enlightening. The even more recent paper of Przymusinski [10] describing nice spaces X with exactly X^n paracompact and X^m normal for reasonable n and m is a meld of techniques. These are not necessarily the latest or most significant papers; they are papers containing useful ideas.

In spite of all the activity, the situation with a number of the most basic old questions is the most frustrating imaginable. We know that, say yes, is consistent with the usual axioms for set theory but we have been unable to find out whether no is consistent ot not:

(1) Is there a p-point in $\beta N-N$? (A point such that the intersection of every countable family of neighborhoods of the point contains a neighborhood.) Such problems involving ultrafilters are set theory in the raw; one hesitates to mention topological formulations. Under a large number of special set theoretic assumptions there are p-points in $\beta N-N$; this is a problem which has had serious attack by

both topologists and logicians . It is probably consistent that there have been no p-points in $\beta N-N$, but no such models have been found. There has been no movement on this problem for several years and in a way the subject has moved on leaving it as an irrelevant island.But i there were a real p-point, the construction would almost surely have many other applications for it is exactly such countable versus uncountable properties which are vital in set theoretic topology.

(2) Is every normal Moore space metrizable?

More generally we have comparison problem: does normal imply collectionwise normal in 1st countable T_2 spaces? Basically nothing is known except that under Martin's axiom plus the continuum hypothesis the answer is no. In translating this problem into set theoretic terms one is given a cardinal λ and one needs to know certain intersection properties for the subsets of subsets of λ . The intractabilit of this problem is based on the fact that the set theoretic questions are 3rd order. The related question: whether normal imples collection wise normal for 1st countable T_2 spaces, is independent of the usua axioms for set theory; here we only need to know about intersection properties for subsets of λ. We have effective tools for dealing with 2^{λ} , but not with $2^{2^{\lambda}}$.

(3) Can a T_3 (1st countable) space have only one of hereditar separability and hereditary Lindelöfness? I suspect this problem is much more set theoretically basic and widely applicable that the previous two problems although people outside of general topology consider the question ridiculously esoteric. Assuming the continuum for hypothesis or the existence of a Souslin line,the answer is yes as a rich variety of examples now show; but the basic question remains. This seems to be a partition calculus problem. $\omega_1 \to (\omega_1; \omega_1)^2$ would imply that the answer is no; and Galvin has shown that this is a consequence of Martin's axiom plus the negation of the continuum hypothesis plus $\omega_1 \to [\omega_1]^2_5$. But $\omega_1 \to [\omega_1]^2_5$ may be false for all we know.

(4) Is there a Dowker space with small cardinal functions?

(Is there a 1st countable or separable or cardinality $\aleph_1$ T_2 space which is normal but not countably paracompact?) There is a normal T space which is not countably paracompact, but to be useful we really need examples which are 1st countable or separable. Consistency examples of such spaces exist . The situation is similar to that of problem (1): we know there are normal T_2 spaces which are not collectionwise normal but we only know that the existence of 1st countable ones is consistent with the usual axioms for set theory. Similarly the existence of a regular hereditarily separable non Lindelöf or 1st countable space in answer to problem (3) would only raise the question of the existence of a 1st countable one. It is typical of set theoretic topology as opposed to set theory that the basic problem is which pathologies are eliminated by which countability conditions.

This year my principal concern has been a problem whose motivation came from Banach spaces. An Eberlein compact is a topological space which is homeomorphic to a weakly compact subset of a Banach spase. (Eberlein proved [11] that a closed subset of a Banach space is compact in the weak topology if and only if it is sequentially compact.) If Γ is a set, let $c_0(\Gamma) = \{f \in I^\Gamma | \{\alpha \in \Gamma | f(\alpha) > \varepsilon\}$ is finite for all $\varepsilon > 0\}$. Amir and Lindenstrauss proved [12] that (for some Γ) every Eberlein compact is homeomorphic to a compact subset of $c_0(\Gamma)$ where the topology on $c_0(\Gamma)$ is just the subspace topology inherited from the product space I^Γ. They conjectured that every (T_2) continuous image of an Eberlein compact is an Eberlein compact. It is.

Rosenthal proved [13] that a compact T_2 space is an Eberlein compact if and only if it has a σ-point-finite point-separating-in-the-T_0-sense family of open F_σ s. (Recall that a compact T_2 space with a point-countable point-separating-in-the-T_1-sense family of open sets is metric). Our solution [14] of the continuous image problem shows, using a good deal of (often finite) combinatories and one straight forward topological lemma, that any T_2 continuous image of a compact subset of $c_0(\Gamma)$ satisfies Rosenthal's characterization of an Eberlein compact.

However, my interest in this problem (which is obviously a topological problem) stemmed from the fact that I thought that the problem was set theoretic. People working on the problem had been asking me questions like: Is every compact T_2 space of cardinality $\leq c$ sequentially compact? Is every compact, T_2 , ccc space with a point-countable point-separating-in-the-T_0-sense family of open F_σ s metrizable? The answer to either of these questions is independent of the usual axioms for set theory. Even those of us who consider ourselves "experts" in set theoretic topology often cannot tell in advance when a problem is indeed set theoretic. The clue to the non set theoretic nature of problems concerning Eberlein compacts is that the weight and cellularity are the same for these spaces and are preserved by irreducible continuous functions; there is little room for cardinal pathology.

A much deeper problem from Banach space theory is: is every injective Banach space isomorphic to $C(S)$ for some extremally disconnected S ? M. Zippin has recently proved by a topological argument that the answer is yes for separable Banach spaces. Perhaps the general problem is purely topological; perhaps set theoretic; it is tempting to try such problems in hopes of finding broader applications for one's set theoretic topological techniques.

I close with a lemma which was the clue to the solution of the Eberlein compact problem mentioned earlier. Neither the lemma nor its proof will come as a surprise to anyone who has proved that metric spaces are paracompact, that locally finite closed refinements yield paracompactness, ...

If λ is an ordinal, we say that a sequence $\{S_\alpha\}_{\alpha<\lambda}$ of subsets of a topological space is left separated provided $S_\alpha \cap \overline{\bigcup_{\beta>\alpha} S_\beta} = \emptyset$ for all $\alpha < \lambda$.

<u>Lemma. If U is an open subset of a T_4 space, then the union of countably many left separated sequences of closed subsets of U covers U.</u>

<u>Proof.</u> Let $\{p_\alpha\}_{\alpha>\lambda}$ be an indexing of the points of U. Fix α. We define $T_{\alpha n}$ for all $n<\omega$ for induction; our induction hypothesis is that $T_{\alpha n}$ is a closed subset of U. Let $T_{\alpha 0} = \{p_\alpha\}$ and, if $T_{\alpha n}$ has been defined, choose a closed set $T_{\alpha,n+1}$ such that $T_{\alpha n} \subset$ interior $T_{\alpha,n+1} \subset U$. Observe that $T_\alpha = \bigcup_{n<\omega} T_{\alpha n}$ is open.

For each $\alpha<\lambda$ and $n \in \omega$, let $S_{\alpha n} = T_{\alpha n} - \bigcup_{n<\omega} T_\beta$. It is trivial that $\{S_{\alpha n}\}_{\alpha<\lambda}$ is a left separated sequence of closed subsets of U. If $p_\alpha \in U$ there is a smallest $\beta \leq \alpha$ such that $p_\alpha \in S_{\beta n}$. So $U = \bigcup_{n<\omega} \bigcup_{\beta<\lambda} \{S_{\beta n}$ and the lemma is proved.

BIBLIOGRAPHY

[1] M.Starbird: The Alexander linear isotopy theorem for dimension 3, (to appear).

[2] R.Shori,J.West: 2^I is homeomorphic to the Hilbert cube,Bull.Amer. Math.Soc.78,no.3(1972), pp.402-406.

[3] M.E.Rudin: Lectures on set theoretic topology, CBMS regional conference series (1974), no.23.

[4] A.J.Ostaszewski: On countably compact, perfectly normal spaces, J.London Math.Soc.

[5] I.Juhász: Cardinal functions in topology, Math.Centre Tracts (1975) Amsterdam, no.34.

[6] B.É.Šapirovskii: Discrete subspaces of topological spaces.Weight, tightness and Souslin number, Dokl.Akad.Nauk SSSR 202 (1972), pp. 779-782 = Soviet Math.Dokl. 13 (1972) no. 1, pp.215-219.

[7] R.Hodel: New proof of a theorem of Hajnal and Juhász on the cardinality of topological spaces, Dept. of Mathematics, Durham,Nort Carolina.

[8] B.A.Efimov: Extermally disconnected compact spaces and absolutes, Trudy Maskov.Mat.Obsc., 23 (1970), pp. 243-285.

[9] J.Chaber:Conditions implying compactness in countably compact spaces. Bull.Acad.Polon.Sci.Sér.Sci.Math.Astronom.Phys.(to appear).

[10] T.Przymusinski: Normality and paracompactness in finite and countable cartesian products,(to appear).

[11] W.Eberlein: Weak compactness in Banach spaces,I.Proc.Acad.Sci. U.S.A., 33 (1947), pp.51-53.

[12] D.Amir,J.Lindenstrauss: The structure of weakly compact sets in Banach spaces, Ann.of Math. 88(1968), pp. 33-46.

[13] H.Rosenthal: The hereditary problem for weakly compactly generated Banach spaces, Comp.Math. 28(1974), pp. 83-111.

[14] J.Benjamini, M.E.Rudin;W.Wage: (to appear).

[15] R.Pol: Short proofs of theorems on cardinality of topological spaces, Pol.Acad.Sci., vol. 22 (1974), no. 2, p. 1245-1249.

SOME TOPOLOGICAL ASPECTS OF THE THEORY OF TOPOLOGICAL TRANSFORMATION GROUPS

Yu. M. Smirnov

Moscow, W-234, M G U , mech. - math.

There are very much interesting and deep results in the classic case, when the subjects for research are topological transformation groups with metrizable compact phase spaces and Lie action groups. The list of all topologists working in this area is very long: A. Borel [1], H. Cartan [2], C. Chevalley [3], A. Gleason [4], J.L. Koszul [5], D. Montgomery [6], G.D. Mostow [7], R. Palais [8], C.T. Yang [9], L. Zippin [10] and others. But I suppose that the general case is waiting yet for its serious studying.

Here we'll speak about some new results of Soviet topologists S. Bogaty [11], M. Madirimov [12], B. Pasynkov [14] and Yu.M. Smirnov [13] in the following directions: DIMENSION, EXTENSION, EMBEDDING for general topological transformation groups.

0. The general transformation group or shortly the G-space is by definition a topological space X (phase space) with a given continuous action $\alpha : G \times X \longrightarrow X$ of a given topological group G (action group). It is convenient to write g(x) instead of $\alpha(g,x)$. A subset A of a given G-space X is called invariant, if $g(A) \subset A$ for every element g of G. A continuous mapping $f: X \longrightarrow Y$ of G-spaces X and Y is called equivariant, if $f(g(x)) = g(f(x))$ for every element g of G and every point x of X.

1. EXTENSION. Let K be a class of topological spaces and G be a fixed group. Then we denote by K(G) the class of all G-spaces for which its phase spaces belong to K. The question is following: Is it possible to characterize the property of a given G-space Y to be an absolute (neighborhood) extensor for a given class K(G) in terms of pure topological properties of some subsets Y!H! of Y ?

Here, naturally, a G-space Y is called an absolute (neighborhood) extensor, briefly AE (resp. ANE) for a given class K(G), if for every G-space X of K(G) and every closed invariant subset A of X, every equivariant mapping $f: A \longrightarrow Y$ has an equivariant extension $F: X \longrightarrow Y$ (resp. $F: U \longrightarrow Y$, where U is an invariant neighborhood of

A in X).

Some necessary conditions are investigated for the following two cases: i) in the "compact case" when the action group G is compact and the class K consists of all compact spaces, ii) in the "metrizable case", when the group G is metrizable and the class K consists of all metrizable spaces. Namely we have the

Theorem 1 S (Smirnov [13 a]). If in these two cases a G-space Y is AE (ANE) for the class K(G) then for every closed subgroup H of G the set Y!H! is AE (resp. ANE) for the class K. Here Y!H! is the set of all "H-fixed" points, i.e. Y!H! = $\{ y \in Y \mid h(y) = y \quad \forall\, h \in H \}$.

This theorem is true also under the following supplementary condition:

AE(n) (resp. ANE(n)): If we take in the definition of the property AE (resp. ANE) for the given topological class K the extensionable mappings f: A $\longrightarrow$ Y!H! , where $A \subset X$, with the property dim $(X \setminus A) \leqq n + 1$ (for a fixed number n), then we shall take in the definition of the property AE (resp. ANE) for the given "G-class" K(G) the extensionable mappings f: A $\longrightarrow$ Y, where $A \subset X$, with the property dim $(X \setminus A) \leqq n + 1 + \dim G$.

Problem 1. Is this theorem true for the "PP-case", when the action group G is a P-paracompact space (in the sense of Arhangelski [15]) and K consists of all P-paracompact spaces ? and for "paracompact case" ??

Problem 2. For which groups G is true the proposition inverse to the proposition of Theorem 1 S ? Is it for compact metrizable zero-dimensional groups ?? Is it for compact metrizable groups ??? and so on ...

The answer is positive for finite commutative action groups, as has been proved by Bogaty:

Theorem 1 B (Bogaty [11 a]). Let the action group G be finite and commutative. Then in metrizable case a finite dimensional G-space Y is AE (resp. ANE) for the class K(G) if and only if for every closed subgroup H of G the set Y!H! is AE (resp. ANE) for the class K.

This theorem is true also under the supplementary condition AE(n) (resp. ANE(n)) without a given restriction of finite dimensionality on the space Y. Then one puts in the proposition of Theorem 1B that the sets Y!H! are LC^n (resp. C^n & LC^n).

Jan Jaworowski [16] has proved this theorem only for the cyclic group Z_p of a prime order p and M. Madirimov [12 a] has established this result for every cyclic group Z_n. The method used by Bogaty-Madirimov is different from the Jaworowski´s method.

2. DIMENSION. There are well known theorems in the dimension theory about some characteristics of the dimension dim X for metrizable spaces X given by M. Katětov [17] and K. Morita [18]. The question is following: Which of these characteristics have some "equivariant" generalizations?

Some of such "equivariant" characteristics of dimension are investigated by M. Madirimov for metrizable phase spaces X and finite action groups G in the following manner. The family ω of sets of a given G-space X is called _equivariant_, if $g(U) \subset U$ for every U of ω and every g of G (Jaworowski [16 a]).

Theorem 1 M (_Madirimov [12 b]_). _For every_ n-_dimensional metrizable_ G-_space_ X _with finite action group_ G

i) _there exist some zero-dimensional metrizable_ G-_space_ M _and some equivariant closed mapping_ $h: M \longrightarrow X$ _onto_ X, _such that the inverse image_ $g^{-1}(x)$ _consists of no more than_ n + 1 _points for every point_ x _of_ X,

ii) _there exist_ n + 1 _zero-dimensional invariant sets_ $U_0, \ldots, U_n$ _such that_ $X = U_0 \cup \ldots \cup U_n$,

iii) _there exists an open basis_ ω _such that_ $\dim(\mathrm{Bd}\ U) \leqq n - 1$ _for every set_ U _of_ ω _and_ ω _is a sum of countably many locally-finite equivariant systems_ ω_i.

Naturally, all inverse propositions are true too by the corresponding topological theorems given by M. Katětov [17] and K. Morita [18].

Corollary. For the orbit space X/G of every metrizable G-space X with a finite action group G the dimension dim X/G = dim X.

Problem 3. For which action groups are true the propositions and Corollary of Theorem 1 M ? Is it for compact metrizable zero-dimensional groups?? Is it for metrizable zero-dimensional groups ??? and so on ...

Problem 4. Is the equality dim X = dim G + dim X/G true for locally-compact phase space X and locally compact action group G, such that all orbits are closed in X ? And if only for compact case??

3. EMBEDDING. Here we have two natural questions: the question about existence of universal G-spaces for some classes of G-spaces and the question about the linearization of some actions on some G-spaces. In this direction there are many interesting papers of P.S. Baayen [19], D.H. Carlson [20], J. de Groot [21], G.D. Mostow [7], R.S. Palais [8 b], J. de Vries [22 ab] and other authors. The last results about universal G-spaces were received I suppose by J. de Vries [22 b] and myself [13 b] independently and almost simultaneously.

Let $L(X)$ be the Lindelöf degree of a topological space X (i.e. the minimum of cardinal numbers n such that every open covering of X has a subcovering of cardinality n).

Theorem 1 V (de Vries [22 b]). For every infinite locally compact group G there exists the G-space $U(G)$ with completely regular phase space such that every completely regular G-space X of weight $w(X)$, where $w(X) \leqq L(G)$, can be topologically and equivariantly embedded into the universal G-space $U(G)$.

It is very interesting his second [22 b]

Theorem 2 V. The G-space X of weight $w(X)$, where $w(X) \leqq L(G)$, is topologically and equivariantly embeddable in some compact G-space Y if and only if the G-space X is bounded (see below). If G is σ-compact and X is separable and metrizable, then Y may be supposed to be metrizable too.

Here the G-space X is said to be bounded, if it is bounded with respect to some uniformity Ω on X (i.e. for every entourage O of Ω there exists a neighborhood U of identity element e of G such that from $g(x) \in U \times G$ follows $(g(x),x) \in O$). +)

I was studying some functorial dependence $\mathfrak{X}$ between the maps h: $X \longrightarrow Y$, where Y is a G-space with an action α, and maps $\mathfrak{X}(h)$: : $X \longrightarrow Y^G$, where Y^G is the space of all maps (continuous or not) from G to Y with some natural action β_G and compact-open topology x).

+) See also the report "Embeddings of G-spaces" given by J. de Vries at this Symposium (Part B of the Proceedings).

x) All "functional" spaces are taken here and below with compact-open topology only!

The "functor" $\tilde{\alpha}$ preserves the following properties of maps h: i) injectivity, ii) continuity, iii) equivariance, if X is a G-space too, iv) a property to be a topological embedding, v) a property to be a closed topological embedding, if the action group G is compact. Here by definition, a topological embedding h is closed, if the image h(X) is closed. In the case ii) $h_\alpha(X) \subset C(G,Y)$, where $h_\alpha = \tilde{\alpha}(h)$ and C(G,Y) is the set of all continuous mappings from G to Y.

The "functor" $\tilde{\alpha}$ is some topological embedding of the space C(X,Y) into the space C(X,C(G,Y)) always supposing the G-space X and the topological space Y are fixed. If Y is a given topological vector space, then the "functor" $\tilde{\alpha}$ is a monomorphism. If G is a locally compact group, then the restriction of the action β_G to $G \times C(G,Y)$ is a continuous action on the group G on the space C(G,Y). Consequently in this case the map $h_\alpha = \tilde{\alpha}(h)$ is an equivariant continuous mapping to C(G,Y) for every continuous mapping h. For any locally convex space Y there exists a continuous monomorphism $\mu_G: G \longrightarrow L(Z)$, where L(Z) is the group of all topological linear transformations of the space Z = C(G,Y).

Theorem 2 S (*Smirnov [13 b]*). *Every completely regular* G*-space* X *with a locally compact action group* G *has some topological equivariant embedding* h_α *into some locally convex space* Z *with the natural action of the group* G *as subgroup of the group* L(Z) (*then this space* Z *is some universal* G*-space). If the action group* G *is compact, then the embedding* h_α *may be supposed to be closed*.

The proof of this theorem is different from the proof of Theorem 1 V given by J. de Vries. I think that the weight condition given by J. de Vries in Theorems 1 V and 2 V can be received in our way too. Almost all our propositions of Theorem 2 S can be illustrated by the following commutative diagram:

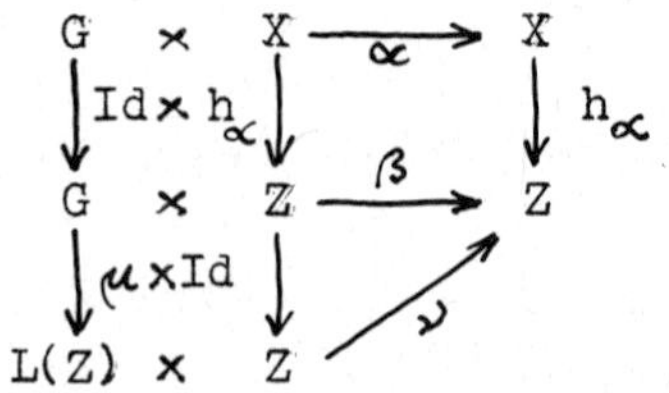

$$\begin{array}{ccc} G \times X & \xrightarrow{\ \alpha\ } & X \\ \downarrow{\scriptstyle Id \times h_\alpha} & & \downarrow{\scriptstyle h_\alpha} \\ G \times Z & \xrightarrow{\ \beta\ } & Z \\ \downarrow{\scriptstyle \mu \times Id} & \nearrow{\scriptstyle \nu} & \\ L(Z) \times Z & & \end{array}$$

Here Z = C(G,Y) and $\nu(p,z) = p(z)$ naturally.

Theorem 2 S is a corollary to one embedding theorem given by R Arens [23] and the following

Theorem 3 S. Every continuous mapping (resp. topological embedding or closed topological embedding) $h: X \longrightarrow Y$ has some corresponding continuous mapping (resp. topological embedding or closed topological embedding always supposing the action group G is compact) $h_\alpha : X \longrightarrow Z$ in some space Z ($= C(G,Y)$) such that for some action $\beta = \beta_G$ on Z and some monomorphism $\mu = \mu_G$ all conditions of Theorem 2 S are satisfied.

The "program" of our proof is the following: we define the maps h_α and $\tilde{\alpha}$, the action β and the monomorphism μ by the formulae $h_\alpha(x)(g) = h(\alpha(g,x))$, $\beta(g',f)(g) = f(g,g')$, $\mu(g)(f) = \beta(g,f)$.

Theorem 4 S. The map $\tilde{\alpha}$ by the hypothesis of Theorem 3 S is a topological isomorphic embedding.

Problem 5. Are Theorems 2 S, 3 S and 4 S true without the condition of local compactness of the action group G ?

4. DIMENSION. Some dimension theorems for topological groups given by B. Pasynkov [14 a] are generalized recently by the same author. Let $\pi : X \longrightarrow X/G$ be the natural projection of a given G-space X on its orbit space X/G. B. Pasynkov calls a completely regular G-space X with the compact action group G almost metrizable if the projection π is a perfect mapping and the orbit space X/G is metrizable.

Theorem 1 P (Pasynkov [14 b]). If the G-space X is almost metrizable, then $\dim X = \operatorname{Ind} X = \Delta X$, where ΔX is the dimension in the sense of V. Ponomarev [24].

Theorem 2 P. Every finite dimensional almost metrizable G-space X has some zero-dimensional perfect mapping on some metric space M, with $\dim M \leqq \dim X$.

To these results is closely related the following

Theorem 3 P. Every almost metrizable compact G-space is dyadic and, moreover, is some Dugundji space [25].

References

1. A. Borel, Transformation groups with two classes of orbits, Proc. Acad. Sci. 43 No 11(Oct. 1957), 983-985.

2. H. Cartan (et al.), Seminar Henri Cartan, 1949-50, Multilith. Paris, 1950.

3. C. Chevalley, On a theorem of Gleason, Proc. Amer. Math. Soc. 2 (No 1)(1951), 122-125.

4. A. Gleason, Spaces with a compact Lie group of transformations, Proc. Amer. Math. Soc. 1(No 1)(Feb. 1950), 35-43.

5. J.L. Coszul, Sur certains groupes de transformation de Lie, Colloque de géométrie différentielle Strasbourg, 1953.

6a. D. Montgomery and L. Zippin, Topological transformation groups, New York, 1955.

6b. D. Montgomery and L. Zippin, A theorem on Lie group, Bull. Amer. Math. Soc. 48(1942), p. 116.

6c. D. Montgomery and C.T. Yang, Differentiable transformation groups on topological spheres, Michigan Math. J. 14(No 1)(1967), 33-46.

7a. G.D. Mostow, Equivariant embedding in euclidean space, Ann. of Math. (65)(No 3)(May 1957), 432-446.

7b. G.D. Mostow, On a conjecture of Montgomery, Ann. of Math.(65) (No 3)(1957), 505-512.

8a. R. Palais, The classification of G-spaces, Mem. Amer. Math. Soc. No 36, 1960.

8b. R. Palais, Embedding of compact differentiable transformation groups in orthogonal representations, J. Math. and Mech. 6(No 5) (Sept. 1957), 673-678.

9. C.T. Yang, On a problem of Montgomery, Proc. Amer. Math. Soc. 8 (No 2)(Aug. 1957), 255-257.

10. L. Zippin, Transformation groups, Lectures in topology, Ann. Arb 1941.

11a. S. Bogaty and M. Madirimov, K teorii razmernosti metričeskich prostranstv s periodičeskimi homeomorfizmami, Mat. Sb. 98(No 1) (1975), 72-83.

12a. M. Madirimov, O razmernosti prostranstva na kotorom dejstvujet gruppa, dep. VINITI, No 1970-75, Jul. 1975.

12b. M. Madirimov, O prodolgeniah ekvivariantnych otobraženij, Mat. Sb. 98(No 1)(1975), 84-92.

13a. Yu. M. Smirnov, Množestvo H-nepodvižnyh toček - absolutnye ekstensory, Mat. Sb. 98(No 1)(1975), 93-101.

13b. Yu.M. Smirnov, Ob ekvivariantnyh vloženijah G-prostranstv, Uspehi Mat. Nauk 31(No 5)(1976), 137-147.

14a. B. Pasynkov, Počti metrizuemye topologičeskie gruppy, Dokl. Akad Nauk SSSR 161 (No 2)(1965), 281-284.

14b. B. Pasynkov, O razmernosti prostranstv s bikompaktnoj gruppoj preobrazovanij, Uspehi Mat. Nauk 31(No 5)(1976), 112-120.

15. A.V. Arhangelskii, Ob odnom klasse prostranstv, soderžatših vse metrizuemye i vse lokalno-bikompaktnye prostranstva, Mat. Sb. 67(No 1)(1965), 55-85.

16a. J.V. Jaworowski, Equivariant extensions of maps, Pacific J.Math. 45(No 1)(1973), 229-244.

16b. J.V. Jaworowski, Extensions of G-maps and euclidean G-retracts, Math.Z. 146(No 2)(1976), 143-148.

17a. M. Katětov, O razmernosti metričeskih prostranstv, Dokl. Akad. Nauk SSSR 79(No 1)(1951), 189-191.

17b. M. Katětov, O razmernosti neseparabelnyh prostranstv I, Czechoslovak Math. J. 2(No 4)(1952), 333-368.

18a. K. Morita, Normal families and dimension theory in metric spaces, Math. Ann. 128(No 4)(1954), 350-362.

18b. K. Morita, A condition for the metrizability of topological spaces and for n-dimensionality, Sci. Rep. Tokyo Kyoiku Daigaku Sect. A,5(1955), 33-36.

19a. P.C. Baayen, Topological linearization of locally compact transformation groups, Report No 2, Wisk. Semin. Free Univ. Amsterdam, 1967.

19b. P.C. Baayen and J. de Groot, Linearization of locally compact transformation groups in Hilbert space, Math. Systems Theory 2 (No 4)(1968), 363-379.

20a. D.H. Carlson, Extensions of dynamical systems via prolongations, Funkcial. Ekvac. 14(No 1)(1971), 35-46.

20b. D.H. Carlson, Universal dynamical systems, Math. Systems Theory 6(No 1)(1972), 90-95.

21. J. de Groot, The action of a locally compact group on a metric space, Nieuw Arch. Wisk.(3)7(No 2)(1959), 70-74.

22a. J. de Vries, A note on topological linearization of locally compact transformation groups in Hilbert space, Math. Systems Theory 6(No 1)(1972), 49-59.

22b. J. de Vries, Universal topological transformation groups, General Topology and Appl. 5(No 2)(1975), 102-122.

22c. J. de Vries, Can every Tychonov G-space equivariant be embedded in a compact G-space? Math. Centrum Amsterdam Afd. Zuivere Wisk. ZW 36, 1975.

23. R. Arens and J. Eells, On embedding uniform and topological spaces, Pacific J. Math. 6(1956), 397-403.

24. V.I. Ponomarev, O nekotoryh primenenijah projekcionnyh spektrov k teorii topologičeskih prostranstv, Dokl.Akad. Nauk SSSR 144

(No 5)(1962), 993-996.

25. A. Pelczinski, Linear extensions, linear averagings and its applications to linear topological classification of spaces of continuous functions, Dissertationes Math. (Rozprawy Mat.) 58, 1968, Warszawa.

MEASURE-PRESERVING MAPS

A. H. Stone

Many important spaces come equipped with measures as well as with topologies. Thus it is of significance to investigate maps that are measure-preserving as well as continuous; and some problems of this nature will be considered here. By a "topological measure space" $(X, \mathcal{T}, \mu)$, or (X, μ) for short, we mean a set X with a topology $\mathcal{T}$ and a countably additive, non-negative regular Borel measure μ, completed with respect to null sets. A map $f: (X, \mathcal{T}, \mu) \to (Y, \mathcal{U}, \nu)$ is "measure-preserving" provided that, for all ν-measurable subsets B of Y, $f^{-1}(B)$ is μ-measurable and $\mu(f^{-1}(B)) = \nu(B)$.

One striking example of a measure-preserving map is Peano's well-known continuous map, say ϕ, of the unit interval I onto the unit square I^2. It is perhaps not well-known that ϕ is measure-preserving when I and I^2 have their usual Lebesgue measures; but this is easily seen by noting that, at the nth stage of the construction of ϕ, I^2 is subdivided into $2^n \times 2^n$ equal squares, the inverse of each of which is an interval in I of the right length (4^{-n}). Recently Schoenfeld [3] has generalized this observation into a measure-preserving form of the Hahn-Mazurkiewicz theorem: if X is a Peano space, with a measure μ such that $\mu(X) = 1$ and μ is positive for every non-empty open subset of X, then there is a continuous measure-preserving map of I (with Lebesgue measure λ) onto X. (Conversely, the conditions on μ here are obviously necessary.)

One of the first major theorems about measure-preserving maps is that of Oxtoby and Ulam [2, p. 886] (see also [5]): if m is a measure on Euclidean space R^n that is non-atomic, positive for all non-empty open sets, and σ-finite but not finite, then there is a measure-preserving homeomorphism from (R^n, m) onto (R^n, λ_n), where λ_n denotes n-dimensional Lebesgue measure. This theorem is deduced from an analogous characterization of Lebesgue measure on I^n: there is a measure-preserving homeomorphism of (I^n, m) onto (I^n, λ_n) if (and only if) m is non-atomic, positive for non-empty open sets, vanishes on the boundary of I^n, and $m(I^n) = 1$. It would be highly desirable to have an extension of this theorem to characterize the product Lebesg measure λ_∞ on the Hilbert cube I^∞; but this seems to be difficult.

A natural question here is: What spaces (X, μ) can be embedded in $(I^\infty, \lambda_\infty)$ by measure-preserving homeomorphisms ? Of course, X must be separable and metrizable, μ must be non-atomic, and $\mu(X)$ must be ≤ 1 The case $\mu(X) = 1$ would imply the extension of the Oxtoby-Ulam theorem just mentioned, so we assume $\mu(X) < 1$. In this form, the question was raised in [4], and answered only in very special cases. The answer is still unknown in general, but the following theorem provides a partial answer that improves on the results stated in [4].

Theorem 1 *If X is a finite-dimensional separable metric space, with a non-atomic complete regular Borel measure μ for which $\mu(X) < 1$, then there is a measure-preserving homeomorphism of (X, μ) onto a subspace of $(I^\infty, \lambda_\infty)$.*

Proof Let $\dim X = n$. Take J to be the interval $[0, 1-\varepsilon] \subset I$, where ε is positive and small enough so $(1-\varepsilon)^{2n+2} > \mu(X)$. First we establish the theorem with $(I^\infty, \lambda_\infty)$ replaced by $(J^{2n+2}, \lambda_{2n+2})$. Take a closed interval K interior to J, and consider $Z = K^{2n+1} \times \{\frac{1}{2}\} \subset J^{2n+2}$. There is a homeomorphism f of X onto a subset Y of Z (see [1, p. 60]). Define a Borel measure m on J^{2n+2} by: $m(B) = \mu(f^{-1}(B \cap Y)) + ((1-\varepsilon)^{2n+2} - \mu(X))\lambda^*_{2n+2}(B - Y)$, where λ^*_{2n+2} is outer Lebesgue measure. Then m (completed with respect to null sets) satisfies the hypotheses of the Oxtoby-Ulam theorem, so that there exists a measure-preserving homeomorphism g of (J^{2n+2}, m) onto $(J^{2n+2}, \lambda_{2n+2})$. Now $g \circ f$ is the measure-preserving embedding of (X, μ) into $(J^{2n+2}, \lambda_{2n+2})$, as required.

We remark, parenthetically, that if X is compact we can replace $2n+2$ by $2n+1$ here, by taking $Y \subset K^{2n+1} \subset J^{2n+1}$. Since Y is closed and n-dimensional it is automatically nowhere dense, so m will still be positive on non-empty open sets.

Next we observe that the interval (J, λ) can be embedded, by a measure-preserving homeomorphism, into $(I^\infty, \lambda_\infty)$. The construction is roughly as follows. It is not hard to see that one can construct a Cantor set C_1 in I^∞ such that $\lambda_\infty(C_1) > 1 - \varepsilon$, and that one can run a simple arc A_1 through C_1. Near each of the countably many complementary intervals of $A_1 - C_1$, we place a small Cantor set of positive measure, and modify A_1 to run through it. Iterating, we end with a simple arc $A \subset I^\infty$ such that $\lambda_\infty(A) > 1 - \varepsilon$ and each sub-arc of A has positive λ_∞-measure. Take a sub-arc A^* of A having $\lambda_\infty(A^*) = 1 - \varepsilon$, and take a homeomorphism h of J onto A^*. For each $t \in J$ put $\phi(t) = \lambda_\infty(h[0, t])$. Then ϕ is a continuous and strictly increasing function, hence a homeomorphism of J onto J; and $h \circ \phi^{-1}$ is a measure-preserving homeomorphism of (J, λ) into $(I^\infty, \lambda_\infty)$, as required.

It follows at once that $(J^{2n+2}, \lambda_{2n+2})$ is imbeddable, by a measure-preserving homeomorphism ψ, into the (topological and measuretheoretic) product $(I^\infty, \lambda_\infty)^{2n+2}$. But this is just $(I^\infty, \lambda_\infty)$. Thus $\theta = \psi \circ g \circ f$ gives a measure-preserving homeomorphism of (X, μ) into $(I^\infty, \lambda_\infty)$, as required.

<u>Remark</u> It cannot be asserted (without extra hypotheses -- for instance, that X is compact, or even analytic) that the images of X in I^{2n+2}, or in I^∞, are Lebesgue measurable. For a measurable subset of positive Lebesgue measure in I^n $(n \leq \infty)$ must contain a Cantor set of positive measure; and this need not be true of (X, μ). Of course, Lebesgue outer measure induces a completed Borel measure on the image of X, whether or not it is measurable; and this is the measure that is "preserved" in Theorem 1.

In a different direction, we have a measure-preserving analogue of Urysohn's Lemma :

<u>Theorem</u> 2 <u>Let</u> X <u>be a topological space, and</u> μ <u>a non-atomic Baire measure on</u> X <u>such that</u> $\mu(X) = 1$. <u>Let</u> F_0, F_1 <u>be disjoint zero-sets in</u> X, <u>both of</u> μ-<u>measure</u> 0. <u>Then there exists a continuous measure-preserving map</u> $g: (X, \mu) \to (I, \lambda)$ <u>such that</u> $g^{-1}(0) \supset F_0$ <u>and</u> $g^{-1}(1) \supset F_1$.

<u>Proof</u> Let $\mathcal{B}$ denote the family of Baire subsets of X. We first note two well-known (and easily proved) facts, the first of which is a consequence of the fact that μ is non-atomic.

1) If $A \in \mathcal{B}$ and $\mu(A) = \alpha > \beta > 0$, there exists $B \in \mathcal{B}$ such that $B \subset A$ and $\mu(B) = \beta$.

(2) Given a zero-set F contained in a cozero-set G, there exists a continuous function $f:X \to I$ such that $F = f^{-1}(0)$ and $X - G = f^{-1}(1)$.

We deduce:

(3) Given a zero-set F contained in a cozero-set G, and $\varepsilon > 0$, there exist a cozero-set U and a zero-set $\tilde{U}$ such that $F \subset U \subset \tilde{U} \subset G$, $\mu(\tilde{U}) < \mu(F) + \varepsilon$, and $\mu(\tilde{U} - U) = 0$.

To prove this, apply (2) and consider the function $g:I \to I$ where $g(t) = \mu(f^{-1}[0, t])$. Then g is a non-decreasing function, hence continuous except for at most countably many values of t. Also g is continuous on the right, hence continuous at 0. Thus there exists $\delta >$ such that $0 \leq t < \delta \Rightarrow g(t) < \mu(F) + \varepsilon$. Choose $t_0 \in (0, \delta]$ to be point of continuity of g, and take $U = f^{-1}[0, t)$, $\tilde{U} = f^{-1}[0, t]$.

In what follows, we continue the same notation: U denotes a cozero-set, and $\tilde{U}$ denotes a zero-set containing U (and hence $\bar{U}$) suc that $\mu(\tilde{U} - U) = 0$.

(4) Given a zero-set F contained in a cozero-set G, where $\mu(G - F)$ $= \alpha > 0$, there exist U, $\tilde{U}$ such that $F \subset U \subset \tilde{U} \subset G$ and $\mu(U - F)$ (and consequently also $\mu(G - \tilde{U})$ is between $\alpha/3$ and 2α

Proof: Take $\varepsilon = \alpha/12$, and apply (3) to get $F \subset U_0 \subset \tilde{U}_0 \subset G$ with $\mu(U_0 - F) < \varepsilon$, and therefore $\alpha - \varepsilon < \mu(G - \tilde{U}_0) \leq \alpha$. From (1) $G - \tilde{U}_0$ contains a Baire set B such that $\mu(B) = \alpha/2$. Since μ (as a finite Baire measure) is automatically regular, there exists a zero-set $Z \subset B$ such that $\mu(B - Z) < \varepsilon$; thus $\alpha/2 - \varepsilon < \mu(Z) \leq \alpha/2$. Applying (3) to Z and $G - \tilde{U}_0$, we get U_1 and $\tilde{U}_1$ such that $Z \subset U_1 \subset \tilde{U}_1 \subset G - \tilde{U}_0$ and $\mu(\tilde{U}_1) < \varepsilon + \mu(Z)$. Put $U = U_0 \cup U_1$, $\tilde{U} = \tilde{U}_0 \cup \tilde{U}_1$; it is easy to verify that the requirements are satisfied.

Now, under the hypotheses of Theorem 2, write $G(0) = \emptyset$, $F(0) = F$ $G(1) = X - F_1$, $F(1) = X$. Applying (4) to $F(0)$ and $G(1)$, we get a cozero-set $G(1/2)$ and a zero-set $F(1/2)$ such that $\mu(F(1/2) - G(1/2)) = 0$, $F(0) \subset G(1/2) \subset F(1/2) \subset G(1)$, and both $\mu(G(1) - F(1/2))$ and $\mu(G(1/2) - F(0))$ are between $1/3$ and $2/3$.

Just as in the proof of the classical Urysohn Lemma, we iterate this procedure, obtaining a system of sets $F(\rho)$, $G(\rho)$, defined for all binary rational numbers ρ in $[0, 1]$, with the following properties

In what follows, it is understood that ρ, σ denote binary rationals in [0, 1].) Then $F(\rho)$ is a zero-set, $G(\rho)$ is a cozero-set, $G(\rho) \subset F(\rho) \subset G(\sigma) \subset F(\sigma)$ whenever $\rho < \sigma$, and $\mu(F(\rho) - G(\rho)) = 0$. Further, in inserting the sets $G((2p+1)/2^{q+1})$ and $F((2p+1)/2^{q+1})$ between $F(p/2^q)$ and $G((p+1)/2^q)$ at the $(q+1)^{st}$ stage, we arrange that both $\mu(G((2p+1)/2^{q+1}) - F(p/2^q))$ and $\mu(G((p+1)/2^q) - F((2p+1)/2^{q+1}))$ are less than $(2/3)^{q+1}$, and both are greater than $(1/3)^{q+1}$.

Now define $f(x) = \sup\{\rho \mid x \notin F(\rho)\}$ for $x \in G(1) - F(0)$. A straightforward verification shows that $f(x) = \inf\{\sigma \mid x \in G(\sigma)\}$. Further, if we define $f(x) = 0$ for $x \in F(0)$, and $f(x) = 1$ for $x \in X - G(1)$, then $f: X \to I$ is continuous, and if $\rho < t < \sigma$, then $f^{-1}(t) \subset G(\sigma) - F(\rho)$. It follows that $\mu(f^{-1}(t)) = 0$ for all $t \in I$, and thus that $\mu(f^{-1}[0, t]) = \mu(f^{-1}[0, t))$.

Finally, define $\phi: I \to I$ by $\phi(t) = \mu(f^{-1}[0, t])$. It follows from the construction that ϕ is a strictly increasing continuous function, and thence that ϕ is a homeomorphism of I onto I. Let $g = \phi \circ f$; it is easy to see that g fulfils all the requirements of the theorem.

Corollary Let $(X, \mathcal{T}, \mu)$ be a topological measure space such that $(X, \mathcal{T})$ is normal, μ is non-atomic, and $\mu(X) = 1$. Let F_0, F_1 be disjoint closed sets in X, both of measure 0. Then there exists a continuous measure-preserving map $g: (X, \mu) \to (I, \lambda)$ such that $g^{-1}(0) \supset F_0$ and $g^{-1}(1) \supset F_1$.

To deduce the Corollary from the theorem, it is enough to show that F_0, F_1, are contained in disjoint zero-sets of measure 0. The regularity of μ gives, for $n = 1, 2, \ldots$, an open set U_n such that $F_0 \subset U_n \subset X - F_1$ and $\mu(U_n) < 1/n$. From the classical Urysohn lemma, there is a continuous function separating F_0 and $X - U_n$, from which we get a zero-set H_n such that $F_0 \subset H_n \subset U_n$. Then $F_0^* = \bigcap_{n=1}^{\infty} H_n$ is a zero-set of measure 0 containing F_0 and disjoint from F_1. Repetition of the argument gives a null zero-set $F_1^* \supset F_1$ and disjoint from F_0^*, as required.

Remark In Theorem 2 (and its Corollary) it would be interesting to know whether one can further arrange that $g(X)$ is a measurable subset of I. The construction used does not in fact ensure this (unless, for instance, X is compact or analytic).

Obviously one could attach to any theorem about the existence of continuous maps the requirement that the maps be measure-preserving, and investigate whether the theorem remains true. We have seen that this is the case (under some restrictions) for Urysohn's Lemma and Urysohn's imbedding theorem. But the "natural" analogue of Tietze's extension theorem is false. For instance, consider the case $(X,\mu) = ($ $A = [0, 1/2]$, and let $f:A \to I$ be defined by $f(x) = 1/2 - x$ (for $0 \leq x \leq 1/2$). Then f is a continuous measure-preserving map of A onto $[0, 1/2]$; but it has no extension to a continuous measure-preserving map $f^*:X \to I$. For the continuity of f^* at $x = 1/2$ would give $\varepsilon > 0$ such that $f^*[1/2, 1/2 + \varepsilon] \subset [0, 1/2)$, and then $f^{*-1}[0, 1/2) \supset [0, 1/2 + \varepsilon]$ and thus has measure $> 1/2$. It would be interesting to have a satisfactory measure-preserving analogue here

Acknowledgement I am grateful to D. Maharam for helpful discussion, and for suggesting Theorem 2.

REFERENCES

[1] W. Hurewicz and H. Wallman, Dimension theory, Princeton 1941.

[2] J. R. Oxtoby and S. M. Ulam, Measure-preserving homeomorphisms and metric transitivity, Ann. of Math. 42 (1941), 874 - 920.

[3] A. H. Schoenfeld, Continuous measure-preserving maps onto Peano spaces, Pacific J. Math. 58 (1975) 627 - 642.

[4] A. H. Stone, Topology and measure theory, Proc. Conference on Measure Theory, Oberwolfach 1975; Lecture Notes in Mathematics No. 641 (Springer-Verlag) .

[5] C. Goffman and G. Pedrick, A proof of the homeomorphism of Lebesgue-Stieltjes Measure with Lebesgue measure, Proc. Amer. Math. Soc. 52 (1975) 196-198.

University of Rochester,
Rochester, N.Y. 14627, U.S.A.

CATEGORIAL ASPECTS ARE USEFUL FOR TOPOLOGY

Věra Trnková
Praha

Under this title, a lecture by M.Hušek and the author was delivered at the Topological symposium. In the lecture, several themes were discussed. We wanted to show some examples how categorial methods and categorial point of view bring or inspire results often "purely topological".

The present paper is a part of this lecture. It consists of two themes discussed in the lecture (the other themes will appear elsewhere), namely

I. EMBEDDINGS OF CATEGORIES

and

II. HOMEOMORPHISMS OF PRODUCTS OF SPACES.

These themes concern distinct fields of problems; however, they are not independent in their methods. The first theme leads e.g. to constructions of stiff classes of spaces (see I.2) and the second one heavily uses them.

The author is indebted to J. Adámek for the reading of some parts of the manuscript and for some comments tending to make the manuscript more lucid.

I.

1. Let us begin with the well-known result of de Groot ([dG]) that every group is isomorphic to the group of all homeomorphisms of a topological space onto itself. In 1964, at the Colloquium on topology in Tihany, he put a problem whether any monoid (i.e. a semigroup with the unit element) is isomorphic to the monoid of all non-constant continuous mappings of a topological space into itself. Let us notice that the set of all non-constant continuous mappings does not always form a monoid, the composition of two non-constant mappings can be constant. The exact formulation is as follows. Given a monoid M, does there exist a space X such that the set of all non-constant continuous mappings of X into itself is closed under composition and this set, endowed with this composition, forms a monoid isomorphic to M ? This was solved positively in [Tr_1], the space X can even be chosen to be metrizable, or, by [Tr_5], compact and Hausdorff. The proof is based on a nice result of Z. Hedrlín and A. Pultr. They proved in [HP] that any small category (i.e. a category, the objects of which form a set) is isomorphic to a full subcategory of the category Graph of all directed graphs and all their compatible mappings. What is really presented in [Tr_1] is a construction of a functor $\mathcal{M}$ of Graph into the category Metr of all metrizable spaces and all their continuous mappings, with the following property. For any pair G, G′ of graphs,

$$f: G \longrightarrow G' \overset{\mathcal{M}}{\rightsquigarrow} \mathcal{M}(f): \mathcal{M}(G) \longrightarrow \mathcal{M}(G')$$

is a bijection of the set of all compatible mappings of G into G′ *onto* the set of all non-constant continuous mappings of $\mathcal{M}(G)$ into $\mathcal{M}(G')$. Since any monoid M can be considered as the set of all morphisms of a category with precisely one object, there exists, by [HP], a graph G such that M is isomorphic to the monoid of all compatible mappings of G into itself. Hence, M is isomorphic to the monoid of all non-constant continuous mappings of $\mathcal{M}(G)$ into itself. In [Tr_5], a functor $\mathcal{C}$ from the category $(\text{Graph})^{op}$, opposite to Graph, into the category Comp of all compact Hausdorff spaces is constructed such that, again, for any pair of graphs, G, G′,

$$f: G \longrightarrow G' \overset{\mathcal{C}}{\rightsquigarrow} \mathcal{C}(f): \mathcal{C}(G') \longrightarrow \mathcal{C}(G)$$

is a bijection of the set of all compatible mappings of G into G′ *onto* the set of all non-constant continuous mappings of $\mathcal{C}(G')$ into $\mathcal{C}(G)$. This makes it possible to obtain the analogous result for compact Hausdorff spaces.

2. These categorial methods give, as a byproduct, some results concerning stiff classes of spaces. Let us recall that a class $\mathbb{C}$ of topological spaces is called stiff if for any X, Y $\in \mathbb{C}$ and any continuous mapping f: X $\longrightarrow$ Y either f is constant or X = Y and f is the identity (sometimes, also the word rigid or strongly rigid is used). Let a cardinal m be given, let k(m) be a discrete category (i.e. with no morphisms except the identities) such that its objects form a set of the cardinality m . Since k(m) is a small category, it is isomorphic to the full subcategory of Graph. Its image under $\mathcal{M}$ is a stiff set (of the cardinality m) of metrizable spaces. Analogously, we obtain arbitratily large stiff sets of compact Hausdorff spaces by means of the functor $\mathcal{C}$. Let us remark that L. Kučera and Z. Hedrlín proved (see [H]) that, under the following set-theoretical assumption

(M) relatively measurable cardinals are not cofinal in the class of all cardinals,

any concrete category is isomorphic to a full subcategory of Graph. A "large discrete category" is concrete, obviously. Consequently, under (M), the functor $\mathcal{M}$ (or $\mathcal{C}$) gives a stiff proper class of metrizable (or compact Hausdorff) spaces. Let us notice that a stiff proper class of paracompact spaces was constructed in [K] without any set-theoretical assumption.

3. Let us recall some usual notions about categories and functors. A functor Φ : K $\longrightarrow$ H is called a full embedding if it is an isomorphism of K onto a full subcategory of H. Now, let H be a category of topological spaces and all their continuous mappings. Φ is called an almost full embedding if, for any pair a, b of objects of K,

$$f: a \longrightarrow b \overset{\Phi}{\rightsquigarrow} \Phi(f): \Phi(a) \longrightarrow \Phi(b)$$

is a bijection of the set of all morphisms of a to b onto the set of all non-constant continuous mappings of Φ(a) to Φ(b). A category U is called universal (or s-universal) if every concrete category (or small category, respectively) can be fully embedded in it. A category T of topological spaces and all their continuous mappings is called almost universal (or almost s-universal) if every concrete category (or small category, respectively) can be almost fully embedded in it. In this terminology, Graph is s-universal and, under (M), it is universal. Metr and Comp are almost s-universal and, under (M), they are almost universal. What V. Koubek really proved in [K] is that the category Par of all paracompact spaces is almost universal. (He starts from a result of L. Kučera and Z. Hedrlín that a rather simply defined category is universal and constructs an almost full embedding of it to the category Par.)

All the above results and their proofs and many others (for example, the investigation of topological categories with other choice of morphisms than all continuous mappings) are contained, with all the details, in the prepared monograph [PT].

4. All the above results say that there are spaces such that all non-constant continuous mappings between any pair of them have some prescribed properties. A classical question of topology is about non-constant continuous mappings into a given space. Let us recall the regular space without non-constant continuous real function of E. Hewitt [Hw] and J. Novák [N] and the following well-known generalization of H. Herrlich [Hr]. For any T_1-space Y there exists a regular space X with more than one point and such that any continuous mapping $f: X \longrightarrow Y$ is constant. Now, we can ask about the coherence of these problems. For example, let a T_1-space Y and a monoid M be given. Does there exist a space X (regular, if possible) such that any continuous mapping $f: X \longrightarrow Y$ is constant and all non-constant continuous mappings of X into itself form a monoid isomorphic to M ? A stronger assertion than the affirmative answer to this question states, is the following

Theorem. For any T_1-space Y, all regular spaces without non-constant continuous mappings into Y (and all their continuous mappings) form an almost universal category.

5. Let us sketch a general construction which gives not only the above theorem but also some further results stated in the next theorems. It is based on the combination of the method used in [Tr_1],[K] and that of [Hr],[EF],[G]. Let Φ_0 be a functor of a category K into the category Top of all topological spaces. Let, for any K-object σ, the space $\Phi_0(\sigma)$ contain a point, say σ_0, with the following property.

For any K-morphism m: $\sigma \longrightarrow \sigma'$, $\Phi_0(m)$ maps $\Phi_0(\sigma) \setminus \{\sigma_0\}$ into $\Phi_0(\sigma') \setminus \{\sigma'_0\}$ and σ_0 on σ'_0.

Let Q be a space with three distinguished points, say q_1, q_2, q_3. By an "iterated glueing" we obtain a new functor $\Psi : K \longrightarrow$ Top. It is defined by induction. We start with Φ_0. If $\Phi_n: K \longrightarrow$ Top and σ_n in every $\Phi_n(\sigma)$ are defined, Φ_{n+1} is obtained as follows. For any

$x \in \Phi_n(\sigma) \setminus \{\sigma_n\}$, we add a copy of Q to $\Phi_n(\sigma)$ and identify q_1 with x, q_2 with σ_n; finally, we identify the q_3's of all copies of Q; the obtained point is σ_{n+1}. If m: $\sigma \longrightarrow \sigma'$ is a K-morphism, then $\Phi_n(m)$ is extended to $\Phi_{n+1}(m)$ so that the copy of Q joining x and σ_n is mapped "identically" onto the copy of Q joining $(\Phi_n(m))(x)$ and σ'_n. Ψ is the union of the functors Φ_n, n = 0,1,2,... .

6. Now, let K be a universal (or s-universal) category and let Ψ be an almost full embedding. Then the range category of Ψ is almost universal (or almost s-universal, respectively) and the spaces $\Psi(\sigma)$ have some desired properties.This is the basic idea of all the proofs. We start with the functor from the universal category into Par, constructed in [K], or from the s-universal category Graph into Metr, constructed in [Tr_1]. This is Φ_0. If Q is suitably chosen, Ψ can be proved to be an almost full embedding.
The construction of Φ_0 in [K] and [Tr_1] as well as the construction of suitable Q heavily use the existence of a Cook continuum [C], i.e. a metrizable continuum H such that, for any subcontinuum L and any continuous mapping f: L $\longrightarrow$ H, either f is constant or f(x) = x for all $x \in L$.

7. We sketch briefly the construction of the space Q. It depends on a given space Z (when Z is a regular totally disconnected space such that any continuous mapping of Z into a given T_1-space Y does not distinguish two points q_1, q_2, we obtain the previous theorem, but other choices of Z are used, too). Let $\mathcal{A}$, $\mathcal{B}_1$, $\mathcal{B}_2$ be countable sets of non-degenerate subcontinua of a Cook continuum H such that $\mathcal{A} \cup \mathcal{B}_1 \cup \mathcal{B}_2$ is pairwise disjoint. $\mathcal{A}$ is used for the construction of Φ_0 as in [K] or [Tr_1]. Then we construct two spaces B_1 and B_2 like in the following figure:

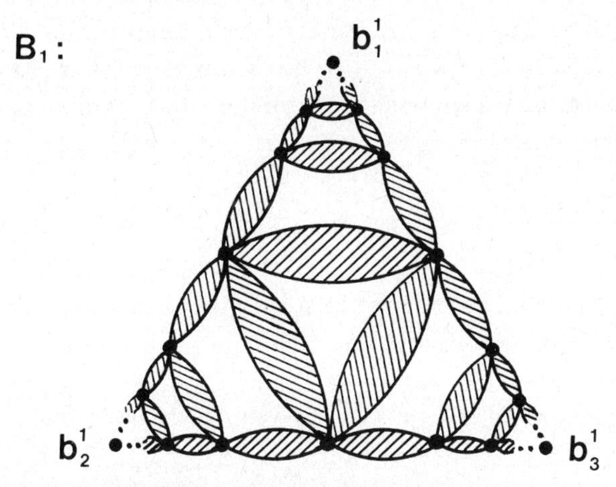

where the [symbol]'s are distinct members of $\mathcal{B}_1$ (in any of them, two distinct points are chosen for the merging). B_2 and b_1^2, b_2^2, b_3^2 are constructed analogously, by means of $\mathcal{B}_2$. We construct Q starting from the space $Z \vee Z'$, where Z is the given space, Z' is a discrete space of the same cardinality, $\vee$ denotes a disjoint union as closed-and-open subsets and $z \longrightarrow z'$ is a bijection of Z onto Z'. For any $z \in Z$ we add a copy of B_1, where we identify b_1^1 with z and b_2^1 with z'. Let $R \subset Z' \times Z'$ be the binary relation described in [VHP]. For any $(y_1, y_2) \in R$, we join y_1 with y_2 by a copy of B_2, i.e. we identify y_1 with b_1^2 and y_2 with b_2^2; finally, we identify all the points b_3^1 and b_3^2 for all the copies of B_1 and B_2. The point obtained by this last identification is q_3, q_1 and q_2 are two distinct points of Z.

8. The theorems stated in 9. and 10. are obtained by this construction if we choose a suitable Z. (The particular choice of Z is always given after the theorem.) In all these cases, it can be seen easily that the spaces $\Psi(\sigma)$ have the required properties. On the other hand, the proof that Ψ is really an almost full embedding, which is the heart of the matter, is more complicated and rather technical.

9. Let V be a topological space. We say that <u>a space X contains V many times</u> if for any $v \in V$ and any $x \in X$ there exists a homeomorphism h of V onto a closed subspace of X such that $h(v) = x$. In the following theorems, speaking about categories of topological spaces, we always mean these spaces and all their continuous mappings. <u>All spaces are supposed to be T_1-spaces.</u>

<u>Theorem.</u> Let V be a paracompact (or normal or completely regular) totally disconnected space. Then all paracompact (or normal or completely regular) spaces, containing V many times, form an almost universal category.

<u>Theorem.</u> Let V be a metrizable totally disconnected space. Then all metrizable spaces, containing V many times, form an almost s-universal category and, under (M), they form an almost universal category.

For the proof of these theorems, we use Z in the above construction as follows. We take a copy of V, say V(v), for any point $v \in V$, and identify all these points v, each in its copy V(v). The obtained point is q_1, $q_2 \in Z \setminus \{q_1\}$ is arbitrary. For the second theorem, all the identifications in the definition of the functor Ψ must be done "metrically".

10. For separation properties weaker than the complete regularity, the construction gives a much stronger result. We can omit the assumption that the given space V is totally disconnected and, simultaneously, continuous mappings in a given space can still be required to be constant. More precisely, the following theorem holds.

Theorem. Let a space Y be given. Let V be a space (or Hausdorff or regular). Then all the spaces (or Hausdorff spaces or regular spaces, respectively) X containing V many times and such that any continuous mapping $f: X \longrightarrow Y$ is constant, form an almost universal category.

For the proof of this theorem, we use Z in the above construction as follows. We take a copy V(v) of V, for any $v \in V$, and identify these points v, as in 9. Denote the obtained space by W, its point obtained by the identifications of the v's by w. Now, let U be a totally disconnected regular space and q_1, q_2 two its distinct points such that, for any continuous mapping f of U into any T_1-space of the cardinality smaller than exp ($\aleph_0 \cdot$ card Y $\cdot$ card W), $f(q_1) = f(q_2)$. (The space constructed in [Hr] or [G] has really only one-point components, q_1 and q_2 are, of course, in one quasicomponent.) Z is a space obtained from a disjoint union of W and U by the identification of w and q_1.

11. Let us show some "purely topological" immediate consequences of the above theorems. By the last theorem,

> there exists a stiff proper class of regular spaces, in which any point lies on an arc.

Another application: since for any set X there exists $R \subset X \times X$ such that the graph (X,R) has no non-identical endomorphism (see [VPH]),

> any totally disconnected space V can be embedded as a closed subspace in a space X without non-constant non-identical continuous mappings into itself such that card X = $2^{\aleph_0}$ $\cdot$ card V and X is completely regular or normal or paracompact or metrizable whenever V has this property.

Analogously,

> for any space V and any cardinal $\alpha >$ card V there exists a space X without non-constant non-identical continuous mappings into itself such that card X = $2^{\aleph_0} \cdot \alpha$, X contains V as a closed subspace and X is Hausdorff or regular whenever V has this property.

Spaces without non-constant non-identical continuous mappings into itself are considered in [KR], where for any infinite cardinal α such Hausdorff space X with card X = α is constructed.

12. The described construction does not "work" for compact spaces. Nevertheless, the following theorem holds.

Theorem. Let V be a totally disconnected compact Hausdorff space. Then all connected compact Hausdorff spaces, containing V many times, form an almost s-universal category. Under (M), they form an almost universal category.

Here, the proof starts from a modification of the almost full embedding of $(\mathrm{Graph})^{op}$ into Comp, described in [Tr_5], and the "iterated glueing" must be done in a different way. The full proof will appear in [Tr_6], where also the proofs of the previous embedding theorems will be given in more detail.

II.

1. In 1957, W. Hanf [H] constructed a Boolean algebra B isomorphic to $B \times B \times B$ but not to $B \times B$. The analogous result for Abelian groups was proved by A.L. Corner in 1963 (see [Cr]). The analogous problem can be investigated in an arbitrary category. Let $\mathbb{K}$ be a category with finite products. Given a natural number $n \geq 3$, denote by

$$\mathbb{K}(n)$$

the class of all objects X of $\mathbb{K}$ such that

X is isomorphic to $X \times \ldots \times X$ (n-times) and $X \times \ldots \times X$ (k-times) is not isomorphic to $X \times \ldots \times X$ (k′-times) whenever $1 \leq k < k' \leq n - 1$.

Let us consider $\mathbb{K}$ to be the category of topological spaces. By [Tr_2], for every n, $\mathbb{K}(n)$ contains a locally compact separable metrizable space. A large part of the method of the proof is categorial, it admits not only a categorial formulation but also an application to other familiar categories. This is done in [Tr_3], where the analogous result is shown also for uniform and proximity spaces, graphs, small categories and some types of partial algebras and unary algebras.

2. Now, we strengthen the above result as follows.

Theorem. Let $\mathbb{K}$ be the category of topological spaces. Let $\mathbb{C}$ be a class of spaces such that

(a) $\mathbb{C}$ contains all metrizable continua;

(b) $\mathbb{C}$ is closed under finite products and countable coproducts (= disjoint unions as clo-open subsets).

Then for any $n \geq 3$ and any X in $\mathbb{C}$ there exist $2^{\aleph_0}$ non-homeomorphic spaces in $\mathbb{C} \cap \mathbb{K}(n)$ such that each of them contains X as a closed subspace and its cardinality is equal to $2^{\aleph_0} \cdot$ card X.

Proof. a) If Y is a space, denote by Y^0 a one-point space, $Y^1 = Y$, $Y^{n+1} = Y \times Y^n$. Denote by N the set of all non-negative integers and by N^N the set of all functions on N with values in N. Let $\{K(x) \mid x \in N \cup \{\infty\}\}$ be a countable stiff set (see I.2) of metrizable continua. For any $\ell \in N^N$ put $K_\ell = \prod_{x \in N} (K(x))^{\ell(x)}$. By [$Tr_3$],

(*) K_ℓ is not homeomorphic to $K_{\ell'}$ whenever $\ell \neq \ell'$.

b) For $\ell, \ell' \in N^N$ define $\ell + \ell'$ by $(\ell + \ell')(x) = \ell(x) + \ell'(x)$. For $A, B \subset N^N$ define $A + B = \{a + b \mid a \in A, b \in B\}$. If n = 1,2,..., put

$n \cdot A = A + \ldots + A$(n-times). Let $n \geq 3$ be given. By [Tr$_3$], there exists a countable set $A \subset N^N$ such that

(i) for any $a \in A$, $a(x) \neq 0$ for infinitely many $x \in N$;

(ii) $A = n \cdot A$;

(iii) if $1 \leq k < k' \leq n - 1$, then $k \cdot A \cap k' \cdot A = \emptyset$.

c) Let a space X in $\mathbb{C}$ be given. Put $Z = X \times K(\infty)$, hence $Z \in \mathbb{C}$. Put

$$\widetilde{Y} = \coprod_{\substack{x \in N \\ a \in A}} (Z^x \times K_a), \qquad \widetilde{Y}_0 = \coprod_{a \in A} (Z^0 \times K_a),$$

where $\coprod$ denotes coproduct. Let Y (or Y_0) be a coproduct of $\aleph_0$ copies of $\widetilde{Y}$ (or $\widetilde{Y}_0$, respectively). Clearly, Y contains X as a closed subspace and card $Y = 2^{\aleph_0} \cdot$ card X. Since Y contains $\aleph_0$ copies of any $Z^x \times K_a$, Y is homeomorphic to Y^n, by (ii).

d) Let us notice that any continuous mapping of $K(\infty)$ into K_ℓ is constant for any $\ell \in N^N$. Hence Y_0 consists precisely of all components C of Y such that any continuous mapping of $K(\infty)$ into C is constant. Consequently, Y_0^k is homeomorphic to $Y_0^{k'}$ whenever Y^k is homeomorphic to $Y^{k'}$, $1 \leq k \leq k' \leq n - 1$. By (iii), this is possible only when $k = k'$. Thus, the space Y has all the required properties.

e) Now, we show that there are many such spaces. Let $\mathfrak{S}$ be a system of infinite subsets of N such that card $\mathfrak{S} = 2^{\aleph_0}$ and, for any distinct $S_1, S_2 \in \mathfrak{S}$, $S_1 \cap S_2$ is finite. Let $\psi_S : N \longrightarrow S$ be a bijection. Construct Y(S) by means of the spaces $\{K(\psi_S(x)) \mid x \in N\}$ quite analogously as Y by means of $\{K(x) \mid x \in N\}$. By (i) and (*), $Y(S_1)$ is not homeomorphic to $Y(S_2)$ whenever S_1 and S_2 are distinct elements of $\mathfrak{S}$.

3. The conditions (a),(b) are not too restrictive, the theorem can be applied e.g. for the class of all spaces, all T_1-spaces, Hausdorff, regular, completely regular, metrizable, σ-compact, realcompact (or E-compact whenever E contains an infinite closed discrete subset and an arc), locally metrizable, spaces with the first or second axiom of countability, separable (or with a density character equal to a given cardinality) and many others. On the other hand, the important class of compact Hausdorff spaces does not satisfy them. Nevertheless, the following theorem holds.

Theorem. For any $n \geq 3$ and any compact Hausdorff (or compact metrizable) space X, there exists a compact Hausdorff (or compact metrizable) space in $\mathbb{K}(n)$ which contains X as a closed subspace.

Outline of the proof. We may suppose that the given space is a cube (or a Hilbert cube). Construct Y analogously as in the previous proof. Let T be a compactification of Y. Since Y is a coproduct of compact connected spaces, $y \in T$ is in Y iff y has a connected neighbourhood in T. Consequently, Y^k is homeomophic to $Y^{k'}$ whenever T^k is homeomorphic to $T^{k'}$ $(1 \leq k \leq k' \leq n-1)$. This is possible only when $k = k'$. Thus, it is sufficient to construct a compactification T of Y such that T is homeomorphic to T^n. The construction will be given in two steps.

a) First, we choose a homeomorphism h of Y onto Y^n and find a compactification T_o of Y such that the following diagram commutes:

$$\begin{array}{ccc} T_o & \xrightarrow{f} & T_o^n \\ \uparrow \iota_o & & \uparrow \iota_o^n \\ Y & \xrightarrow[h]{} & Y^n \end{array} ,$$

where ι_o is the embedding and f is a continuous mapping (if $g: P \longrightarrow Q$ is a mapping, we denote by $g^n: P^n \longrightarrow Q^n$ the mapping defined by $g^n(p_1,\ldots,p_n) = (g(p_1),\ldots,g(p_n))$). This is easy for compact Hausdorff spaces; we put $T_o = \beta Y$ and f is a continuous extension of $\iota_o^n \circ h$. Now, we construct a metrizable compactification T_o for metrizable Y. Let H be the Hilbert cube, $\varkappa : Y \longrightarrow H$ an embedding. Define $\lambda : Y \longrightarrow \prod_{k=0}^{\infty} H^{n^k}$ by

$$\lambda(y) = (\varkappa(y), \varkappa^n(h(y)), \varkappa^{n^2}(h^n(y)),\ldots, \varkappa^{n^{k+1}}(h^{n^k}(y)),\ldots)$$

and $\tau : \prod_{k=0}^{\infty} H^{n^k} \longrightarrow \prod_{k=1}^{\infty} H^{n^k}$ by $\tau(z_0,z_1,z_2,\ldots) = (z_1,z_2,\ldots)$, where $z_i \in H^{n^i}$. Denote $R = \{1,2,\ldots,n\}$. Then any $z \in (\prod_{k=0}^{\infty} H^{n^k})^n$ can be expressed as $z = (((z_{i,j,k} \mid j \in R^k) \mid k = 0,1,2,\ldots) \mid i \in R)$. Define a homeomorphism $\sigma : (\prod_{k=0}^{\infty} H^{n^k})^n \longrightarrow \prod_{k=1}^{\infty} H^{n^k}$ by $\sigma(z) = ((z_{i,j,k} \mid (i,j) \in R \times R^k) \mid k = 0,1,2,\ldots)$. One can verify that $(\sigma^{-1} \circ \tau) \circ \lambda = \lambda^n \circ h$. Then define T_o as the closure of $\lambda(Y)$ in $\prod_{k=0}^{\infty} H^{n^k}$ and f as the corresponding domain-range-restriction of $\sigma^{-1} \circ \tau$.

b) Now, we consider the following diagram.

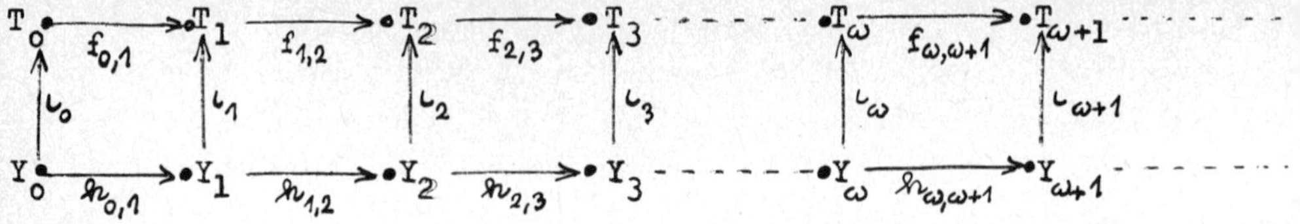

$$\begin{array}{ccccccccccc}
T_0 & \xrightarrow{f_{0,1}} & T_1 & \xrightarrow{f_{1,2}} & T_2 & \xrightarrow{f_{2,3}} & T_3 & \cdots & T_\omega & \xrightarrow{f_{\omega,\omega+1}} & T_{\omega+1} \cdots \\
\uparrow \iota_0 & & \uparrow \iota_1 & & \uparrow \iota_2 & & \uparrow \iota_3 & & \uparrow \iota_\omega & & \uparrow \iota_{\omega+1} \\
Y_0 & \xrightarrow{h_{0,1}} & Y_1 & \xrightarrow{h_{1,2}} & Y_2 & \xrightarrow{h_{2,3}} & Y_3 & \cdots & Y_\omega & \xrightarrow{h_{\omega,\omega+1}} & Y_{\omega+1} \cdots
\end{array}$$

where $Y_o = Y$, $h_{o,1} = h$, $f_{o,1} = f$ are as in a) and $T_{i+1} = T_i^n$, $\iota_{i+1} = \iota_i^n$, $h_{i+1,i+2} = (h_{i,i+1})^n$, $f_{i+1,i+2} = (f_{i,i+1})^n$, Y_ω and T_ω are colimits in the category of all Hausdorff spaces of the preceding chains (hence, T_ω is a compact Hausdorff space; it is metrizable whenever T_o is metrizable). The proof that ι_ω is a homeomorphism of Y_ω into T_ω is omitted as well as the definition of $h_{\omega,\omega+1}$ and $f_{\omega,\omega+1}$ whenever ω is a limit ordinal (this definition is "natural", use the fact that $Y_{i+1} = Y_i^n$, $T_{i+1} = T_i^n$). All $h_{\alpha,\beta}$ are homeomorphisms of Y_α onto Y_β, all $f_{\alpha,\beta}$ are surjective continuous mappings. Since all the T_α's are quotients of T_o, this process must stop, i.e. $f_{\alpha,\alpha+1}$ must be a homeomorphism for some ordinal α. Then T_α is a compactification with the required properties.

4. The proofs of all the above theorems are based on the stiff set $\{K(x) \mid x \in N\}$ of non-degenerate continua. Thus, none of the constructed spaces is zero-dimensional. Nevertheless, the following theorem holds (the proof will appear in [TK]).

Theorem. For any $n \geq 3$, any Boolean space can be embedded into a Boolean space from $\mathbb{K}(n)$.

5. Let us sketch a more general setting of the above field of problems. Let $\mathbb{K}$ be a category with finite products, let $(S,+)$ be a commutative semigroup. Any mapping

$$r: S \longrightarrow \mathrm{obj}\, \mathbb{K}$$

is called a representation of the semigroup by products in $\mathbb{K}$ provided that for any $s_1, s_2 \in S$, $r(s_1 + s_2)$ is isomorphic to $r(s_1) \times r(s_2)$ and $r(s_1)$ is not isomorphic to $r(s_2)$ whenever $s_1 \neq s_2$.
Hence, any object X from $\mathbb{K}(n)$ generates a representation of the finite cyclic group of the order $n-1$. In $[Tr_3]$, a general method is described for the representation of any semigroup $\exp N^M$ (here, N^M is the semigroup of all functions on M with values in N, $\exp N^M$ is the semigroup of all its subsets) in several familiar categories, including the category of topological or uniform or proximity spaces. By $[Tr_4]$, any commutative semigroup S can be embedded into $\exp N^{\aleph_0 \cdot \mathrm{card}\, S}$

Hence, any commutative semigroup S has a representation by products of topological spaces. These spaces can be chosen to be coproducts of continua, by [Tr_5], or coproducts of Boolean spaces, by [AK]. The results presented in I. of this paper imply the following assertions as an immediate consequence.

Given a T_1-space X (or Hausdorff or regular), any commutative semigroup has a representation by products of T_1-spaces (or Hausdorff or regular) containing X many times.

Given a totally disconnected Tichonov space X, any commutative semigroup has a representation by products of Tichonov spaces containing X many times.

The method, used in II.2, can be used to prove easily the following assertion.

Let $\mathfrak{C}$ be a class of spaces containing all continua and closed under finite products and arbitrary coproducts. Let X be a space, let $\mathfrak{C}(X)$ be the class of all spaces from $\mathfrak{C}$, which contain X as a closed subspace. If $\mathfrak{C}(X) \neq \emptyset$, then any commutative semigroup has a representation by products of spaces from the $\mathfrak{C}(X)$.

In [AK], the semigroups $(\exp N^N)^M$ are represented by products such that only countable products of spaces of the basic system $\{K(x) \mid x \in N \times M\}$ are used. This makes it possible to represent the class of all semigroups, embeddable in $(\exp N^N)^M$ for some set M, by products of metrizable spaces. This class of semigroups contains all Abelian groups. Thus, the method of [AK] and the results presented here can be used to prove easily the following assertions.

Given a totally disconnected metrizable space X, any Abelian group has a representation by products of metrizable spaces containing X many times.

Let $\mathfrak{C}$ be a class of spaces containing all complete metric semicontinua and closed under finite products and arbitrary coproducts. Let X be a space, $\mathfrak{C}(X)$ the class of all spaces from $\mathfrak{C}$ which contain X as a closed subspace. If $\mathfrak{C}(X) \neq \emptyset$, then any Abelian group has a representation by products of spaces from $\mathfrak{C}(X)$.

References

[AK] J, Adámek, V. Koubek, On representations of ordered commutative semigroups, to appear.

[C] H. Cook, Continua which admit only the identity mapping onto non-degenerate subcontinua, Fund. Math. 60(1967), 241-249.

[Cr] A.L. Corner, On a conjecture of Pierce concerning direct decomposition of Abelian group, Proc. of Coll. on Abelian groups, Tihany 1963, 43-48.

[EF] W.T. van Est and H. Freudenthal, Trennung durch stetige Funktionen in topologischen Räumen, Indagationes Math. 13(1951), 359-368.

[G] T.E. Gantner, A regular space on which every continuous real-valued function is constant, Amer. Math. Monthly 78(1971), 52-53.

[dG] J. de Groot, Groups represented by homeomorphism groups I., Math. Ann. 138(1959), 80-102.

[H] W. Hanf, On some fundamental problems concerning isomorphism of Boolean algebras, Math. Scand. 5(1957), 205-217.

[Hd] Z. Hedrlín, Extension of structures and full embeddings of categories, Actes de Congrès International des Mathématiciens 1970, tome 1, Paris 1971, 319-322.

[HP] Z. Hedrlín, A. Pultr, O predstavlenii malych kategorij, Dokl. AN SSSR 160(1965), 284-286.

[Hr] H. Herrlich, Wann sind alle stetigen Abbildungen in Y konstant? Math. Zeitschr. 90(1965), 152-154.

[Hw] E. Hewitt, On two problems of Urysohn, Ann. Math. 47(1946), 503-509.

[KR] V. Kannan, M. Rajagopalan, Constructions and applications of rigid spaces I. (preprint).

[K] V. Koubek, Each concrete category has a representation by T_2-paracompact topological spaces, Comment. Math. Univ. Carolinae 15(1974), 655-663.

[N] J. Novák, Regular space on which every continuous function is constant, Časopis pro pěst. mat. fys. 73(1948), 58-68.

[PT] A. Pultr, V. Trnková, Combinatorial, algebraic and topological representations of groups, semigroups and categories, to appear

[Tr_1] V. Trnková, Non-constant continuous mappings of metric or compact Hausdorff spaces, Comment. Math. Univ. Carolinae 13(1972), 283-295.

[Tr_2] V. Trnková, X^n is homeomorphic to X^m iff $n \sim m$, where $\sim$ is a congruence on natural numbers, Fund. Math. 80(1973), 51-56.

[Tr_3] V. Trnková, Representation of semigroups by products in a category, J. Algebra 34(1975), 191-204.

[Tr_4] V. Trnková, On a representation of commutative semigroups, Semigroup Forum 10(1975), 203-214.

[Tr_5] V. Trnková, Vsje malyje kategorii predstavimy nepreryvnymi nepostojannymi otobraženijami bikompaktov, Doklady AN SSSR 230 (1976), 789-791.

[Tr_6] V. Trnková, Topological spaces with prescribed non-constant continuous mappings, to appear.

[TK] V. Trnková, V. Koubek, Isomorphisms of sums of Boolean algebras, to appear.

[VPH] P. Vopěnka, A. Pultr, Z. Hedrlín, A rigid relation exists on any set, Comment. Math. Univ. Carolinae 6(1965), 149-155.

Vol. 460: O. Loos, Jordan Pairs. XVI, 218 pages. 1975.

Vol. 461: Computational Mechanics. Proceedings 1974. Edited by J. T. Oden. VII, 328 pages. 1975.

Vol. 462: P. Gérardin, Construction de Séries Discrètes p-adiques. »Sur les séries discrètes non ramifiées des groupes réductifs déployés p-adiques«. III, 180 pages. 1975.

Vol. 463: H.-H. Kuo, Gaussian Measures in Banach Spaces. VI, 224 pages. 1975.

Vol. 464: C. Rockland, Hypoellipticity and Eigenvalue Asymptotics. III, 171 pages. 1975.

Vol. 465: Séminaire de Probabilités IX. Proceedings 1973/74. Edité par P. A. Meyer. IV, 589 pages. 1975.

Vol. 466: Non-Commutative Harmonic Analysis. Proceedings 1974. Edited by J. Carmona, J. Dixmier and M. Vergne. VI, 231 pages. 1975.

Vol. 467: M. R. Essén, The Cos $\pi\lambda$ Theorem. With a paper by Christer Borell. VII, 112 pages. 1975.

Vol. 468: Dynamical Systems – Warwick 1974. Proceedings 1973/74. Edited by A. Manning. X, 405 pages. 1975.

Vol. 469: E. Binz, Continuous Convergence on C(X). IX, 140 pages. 1975.

Vol. 470: R. Bowen, Equilibrium States and the Ergodic Theory of Anosov Diffeomorphisms. III, 108 pages. 1975.

Vol. 471: R. S. Hamilton, Harmonic Maps of Manifolds with Boundary. III, 168 pages. 1975.

Vol. 472: Probability-Winter School. Proceedings 1975. Edited by Z. Ciesielski, K. Urbanik, and W. A. Woyczyński. VI, 283 pages. 1975.

Vol. 473: D. Burghelea, R. Lashof, and M. Rothenberg, Groups of Automorphisms of Manifolds. (with an appendix by E. Pedersen) VII, 156 pages. 1975.

Vol. 474: Séminaire Pierre Lelong (Analyse) Année 1973/74. Edité par P. Lelong. VI, 182 pages. 1975.

Vol. 475: Répartition Modulo 1. Actes du Colloque de Marseille-Luminy, 4 au 7 Juin 1974. Edité par G. Rauzy. V, 258 pages. 1975. 1975.

Vol. 476: Modular Functions of One Variable IV. Proceedings 1972. Edited by B. J. Birch and W. Kuyk. V, 151 pages. 1975.

Vol. 477: Optimization and Optimal Control. Proceedings 1974. Edited by R. Bulirsch, W. Oettli, and J. Stoer. VII, 294 pages. 1975.

Vol. 478: G. Schober, Univalent Functions – Selected Topics. V, 200 pages. 1975.

Vol. 479: S. D. Fisher and J. W. Jerome, Minimum Norm Extremals in Function Spaces. With Applications to Classical and Modern Analysis. VIII, 209 pages. 1975.

Vol. 480: X. M. Fernique, J. P. Conze et J. Gani, Ecole d'Eté de Probabilités de Saint-Flour IV-1974. Edité par P.-L. Hennequin. XI, 293 pages. 1975.

Vol. 481: M. de Guzmán, Differentiation of Integrals in R^n. XII, 226 pages. 1975.

Vol. 482: Fonctions de Plusieurs Variables Complexes II. Séminaire François Norguet 1974–1975. IX, 367 pages. 1975.

Vol. 483: R. D. M. Accola, Riemann Surfaces, Theta Functions, and Abelian Automorphisms Groups. III, 105 pages. 1975.

Vol. 484: Differential Topology and Geometry. Proceedings 1974. Edited by G. P. Joubert, R. P. Moussu, and R. H. Roussarie. IX, 287 pages. 1975.

Vol. 485: J. Diestel, Geometry of Banach Spaces – Selected Topics. XI, 282 pages. 1975.

Vol. 486: S. Stratila and D. Voiculescu, Representations of AF-Algebras and of the Group U (∞). IX, 169 pages. 1975.

Vol. 487: H. M. Reimann und T. Rychener, Funktionen beschränkter mittlerer Oszillation. VI, 141 Seiten. 1975.

Vol. 488: Representations of Algebras, Ottawa 1974. Proceedings 1974. Edited by V. Dlab and P. Gabriel. XII, 378 pages. 1975.

Vol. 489: J. Bair and R. Fourneau, Etude Géométrique des Espaces Vectoriels. Une Introduction. VII, 185 pages. 1975.

Vol. 490: The Geometry of Metric and Linear Spaces. Proceedings 1974. Edited by L. M. Kelly. X, 244 pages. 1975.

Vol. 491: K. A. Broughan, Invariants for Real-Generated Uniform Topological and Algebraic Categories. X, 197 pages. 1975.

Vol. 492: Infinitary Logic: In Memoriam Carol Karp. Edited by D. W. Kueker. VI, 206 pages. 1975.

Vol. 493: F. W. Kamber and P. Tondeur, Foliated Bundles and Characteristic Classes. XIII, 208 pages. 1975.

Vol. 494: A Cornea and G. Licea. Order and Potential Resolvent Families of Kernels. IV, 154 pages. 1975.

Vol. 495: A. Kerber, Representations of Permutation Groups II. V, 175 pages. 1975.

Vol. 496: L. H. Hodgkin and V. P. Snaith, Topics in K-Theory. Two Independent Contributions. III, 294 pages. 1975.

Vol. 497: Analyse Harmonique sur les Groupes de Lie. Proceedings 1973–75. Edité par P. Eymard et al. VI, 710 pages. 1975.

Vol. 498: Model Theory and Algebra. A Memorial Tribute to Abraham Robinson. Edited by D. H. Saracino and V. B. Weispfenning. X, 463 pages. 1975.

Vol. 499: Logic Conference, Kiel 1974. Proceedings. Edited by G. H. Müller, A. Oberschelp, and K. Potthoff. V, 651 pages 1975.

Vol. 500: Proof Theory Symposion, Kiel 1974. Proceedings. Edited by J. Diller and G. H. Müller. VIII, 383 pages. 1975.

Vol. 501: Spline Functions, Karlsruhe 1975. Proceedings. Edited by K. Böhmer, G. Meinardus, and W. Schempp. VI, 421 pages. 1976.

Vol. 502: János Galambos, Representations of Real Numbers by Infinite Series. VI, 146 pages. 1976.

Vol. 503: Applications of Methods of Functional Analysis to Problems in Mechanics. Proceedings 1975. Edited by P. Germain and B. Nayroles. XIX, 531 pages. 1976.

Vol. 504: S. Lang and H. F. Trotter, Frobenius Distributions in GL_2-Extensions. III, 274 pages. 1976.

Vol. 505: Advances in Complex Function Theory. Proceedings 1973/74. Edited by W. E. Kirwan and L. Zalcman. VIII, 203 pages. 1976.

Vol. 506: Numerical Analysis, Dundee 1975. Proceedings. Edited by G. A. Watson. X, 201 pages. 1976.

Vol. 507: M. C. Reed, Abstract Non-Linear Wave Equations. VI, 128 pages. 1976.

Vol. 508: E. Seneta, Regularly Varying Functions. V, 112 pages. 1976.

Vol. 509: D. E. Blair, Contact Manifolds in Riemannian Geometry. VI, 146 pages. 1976.

Vol. 510: V. Poènaru, Singularités C^∞ en Présence de Symétrie. V, 174 pages. 1976.

Vol. 511: Séminaire de Probabilités X. Proceedings 1974/75. Edité par P. A. Meyer. VI, 593 pages. 1976.

Vol. 512: Spaces of Analytic Functions, Kristiansand, Norway 1975. Proceedings. Edited by O. B. Bekken, B. K. Øksendal, and A. Stray. VIII, 204 pages. 1976.

Vol. 513: R. B. Warfield, Jr. Nilpotent Groups. VIII, 115 pages. 1976.

Vol. 514: Séminaire Bourbaki vol. 1974/75. Exposés 453 – 470. IV, 276 pages. 1976.

Vol. 515: Bäcklund Transformations. Nashville, Tennessee 1974. Proceedings. Edited by R. M. Miura. VIII, 295 pages. 1976.

Vol. 516: M. L. Silverstein, Boundary Theory for Symmetric Markov Processes. XVI, 314 pages. 1976.

Vol. 517: S. Glasner, Proximal Flows. VIII, 153 pages. 1976.

Vol. 518: Séminaire de Théorie du Potentiel, Proceedings Paris 1972–1974. Edité par F. Hirsch et G. Mokobodzki. VI, 275 pages. 1976.

Vol. 519: J. Schmets, Espaces de Fonctions Continues. XII, 150 pages. 1976.

Vol. 520: R. H. Farrell, Techniques of Multivariate Calculation. X, 337 pages. 1976.

Vol. 521: G. Cherlin, Model Theoretic Algebra – Selected Topics. IV, 234 pages. 1976.

Vol. 522: C. O. Bloom and N. D. Kazarinoff, Short Wave Radiation Problems in Inhomogeneous Media: Asymptotic Solutions. V. 104 pages. 1976.

Vol. 523: S. A. Albeverio and R. J. Høegh-Krohn, Mathematical Theory of Feynman Path Integrals. IV, 139 pages. 1976.

Vol. 524: Séminaire Pierre Lelong (Analyse) Année 1974/75. Edité par P. Lelong. V, 222 pages. 1976.

Vol. 525: Structural Stability, the Theory of Catastrophes, and Applications in the Sciences. Proceedings 1975. Edited by P. Hilton. VI, 408 pages. 1976.

Vol. 526: Probability in Banach Spaces. Proceedings 1975. Edited by A. Beck. VI, 290 pages. 1976.

Vol. 527: M. Denker, Ch. Grillenberger, and K. Sigmund, Ergodic Theory on Compact Spaces. IV, 360 pages. 1976.

Vol. 528: J. E. Humphreys, Ordinary and Modular Representations of Chevalley Groups. III, 127 pages. 1976.

Vol. 529: J. Grandell, Doubly Stochastic Poisson Processes. X, 234 pages. 1976.

Vol. 530: S. S. Gelbart, Weil's Representation and the Spectrum of the Metaplectic Group. VII, 140 pages. 1976.

Vol. 531: Y.-C. Wong, The Topology of Uniform Convergence on Order-Bounded Sets. VI, 163 pages. 1976.

Vol. 532: Théorie Ergodique. Proceedings 1973/1974. Edité par J.-P. Conze and M. S. Keane. VIII, 227 pages. 1976.

Vol. 533: F. R. Cohen, T. J. Lada, and J. P. May, The Homology of Iterated Loop Spaces. IX, 490 pages. 1976.

Vol. 534: C. Preston, Random Fields. V, 200 pages. 1976.

Vol. 535: Singularités d'Applications Differentiables. Plans-sur-Bex. 1975. Edité par O. Burlet et F. Ronga. V, 253 pages. 1976.

Vol. 536: W. M. Schmidt, Equations over Finite Fields. An Elementary Approach. IX, 267 pages. 1976.

Vol. 537: Set Theory and Hierarchy Theory. Bierutowice, Poland 1975. A Memorial Tribute to Andrzej Mostowski. Edited by W. Marek, M. Srebrny and A. Zarach. XIII, 345 pages. 1976.

Vol. 538: G. Fischer, Complex Analytic Geometry. VII, 201 pages. 1976.

Vol. 539: A. Badrikian, J. F. C. Kingman et J. Kuelbs, Ecole d'Eté de Probabilités de Saint Flour V-1975. Edité par P.-L. Hennequin. IX, 314 pages. 1976.

Vol. 540: Categorical Topology, Proceedings 1975. Edited by E. Binz and H. Herrlich. XV, 719 pages. 1976.

Vol. 541: Measure Theory, Oberwolfach 1975. Proceedings. Edited by A. Bellow and D. Kölzow. XIV, 430 pages. 1976.

Vol. 542: D. A. Edwards and H. M. Hastings, Čech and Steenrod Homotopy Theories with Applications to Geometric Topology. VII, 296 pages. 1976.

Vol. 543: Nonlinear Operators and the Calculus of Variations, Bruxelles 1975. Edited by J. P. Gossez, E. J. Lami Dozo, J. Mawhin, and L. Waelbroeck, VII, 237 pages. 1976.

Vol. 544: Robert P. Langlands, On the Functional Equations Satisfied by Eisenstein Series. VII, 337 pages. 1976.

Vol. 545: Noncommutative Ring Theory. Kent State 1975. Edited by J. H. Cozzens and F. L. Sandomierski. V, 212 pages. 1976.

Vol. 546: K. Mahler, Lectures on Transcendental Numbers. Edited and Completed by B. Diviš and W. J. Le Veque. XXI, 254 pages. 1976.

Vol. 547: A. Mukherjea and N. A. Tserpes, Measures on Topological Semigroups: Convolution Products and Random Walks. V, 197 pages. 1976.

Vol. 548: D. A. Hejhal, The Selberg Trace Formula for PSL (2,ℝ). Volume I. VI, 516 pages. 1976.

Vol. 549: Brauer Groups, Evanston 1975. Proceedings. Edited by D. Zelinsky. V, 187 pages. 1976.

Vol. 550: Proceedings of the Third Japan – USSR Symposium on Probability Theory. Edited by G. Maruyama and J. V. Prokhorov. VI, 722 pages. 1976.

Vol. 551: Algebraic K-Theory, Evanston 1976. Proceedings. Edited by M. R. Stein. XI, 409 pages. 1976.

Vol. 552: C. G. Gibson, K. Wirthmüller, A. A. du Plessis and E. J. N. Looijenga. Topological Stability of Smooth Mappings. V, 155 pages. 1976.

Vol. 553: M. Petrich, Categories of Algebraic Systems. Vector and Projective Spaces, Semigroups, Rings and Lattices. VIII, 217 pages. 1976.

Vol. 554: J. D. H. Smith, Mal'cev Varieties. VIII, 158 pages. 1976.

Vol. 555: M. Ishida, The Genus Fields of Algebraic Number Fields. VII, 116 pages. 1976.

Vol. 556: Approximation Theory. Bonn 1976. Proceedings. Edited by R. Schaback and K. Scherer. VII, 466 pages. 1976.

Vol. 557: W. Iberkleid and T. Petrie, Smooth S^1 Manifolds. III, 163 pages. 1976.

Vol. 558: B. Weisfeiler, On Construction and Identification of Graphs. XIV, 237 pages. 1976.

Vol. 559: J.-P. Caubet, Le Mouvement Brownien Relativiste. IX, 212 pages. 1976.

Vol. 560: Combinatorial Mathematics, IV, Proceedings 1975. Edited by L. R. A. Casse and W. D. Wallis. VII, 249 pages. 1976.

Vol. 561: Function Theoretic Methods for Partial Differential Equations. Darmstadt 1976. Proceedings. Edited by V. E. Meister, N. Weck and W. L. Wendland. XVIII, 520 pages. 1976.

Vol. 562: R. W. Goodman, Nilpotent Lie Groups: Structure and Applications to Analysis. X, 210 pages. 1976.

Vol. 563: Séminaire de Théorie du Potentiel. Paris, No. 2. Proceedings 1975–1976. Edited by F. Hirsch and G. Mokobodzki. VI, 292 pages. 1976.

Vol. 564: Ordinary and Partial Differential Equations, Dundee 1976. Proceedings. Edited by W. N. Everitt and B. D. Sleeman. XVIII, 551 pages. 1976.

Vol. 565: Turbulence and Navier Stokes Equations. Proceedings 1975. Edited by R. Temam. IX, 194 pages. 1976.

Vol. 566: Empirical Distributions and Processes. Oberwolfach 1976. Proceedings. Edited by P. Gaenssler and P. Révész. VII, 146 pages. 1976.

Vol. 567: Séminaire Bourbaki vol. 1975/76. Exposés 471–488. IV, 303 pages. 1977.

Vol. 568: R. E. Gaines and J. L. Mawhin, Coincidence Degree, and Nonlinear Differential Equations. V, 262 pages. 1977.

Vol. 569: Cohomologie Etale SGA 4½. Séminaire de Géométrie Algébrique du Bois-Marie. Edité par P. Deligne. V, 312 pages. 1977.

Vol. 570: Differential Geometrical Methods in Mathematical Physics, Bonn 1975. Proceedings. Edited by K. Bleuler and A. Reetz. VIII, 576 pages. 1977.

Vol. 571: Constructive Theory of Functions of Several Variables, Oberwolfach 1976. Proceedings. Edited by W. Schempp and K. Zeller. VI, 290 pages. 1977

Vol. 572: Sparse Matrix Techniques, Copenhagen 1976. Edited by V. A. Barker. V, 184 pages. 1977.

Vol. 573: Group Theory, Canberra 1975. Proceedings. Edited by R. A. Bryce, J. Cossey and M. F. Newman. VII, 146 pages. 1977.

Vol. 574: J. Moldestad, Computations in Higher Types. IV, 203 pages. 1977.

Vol. 575: K-Theory and Operator Algebras, Athens, Georgia 1975. Edited by B. B. Morrel and I. M. Singer. VI, 191 pages. 1977.

Vol. 576: V. S. Varadarajan, Harmonic Analysis on Real Reductive Groups. VI, 521 pages. 1977.

Vol. 577: J. P. May, E_∞ Ring Spaces and E_∞ Ring Spectra. IV, 268 pages. 1977.

Vol. 578: Séminaire Pierre Lelong (Analyse) Année 1975/76. Edité par P. Lelong. VI, 327 pages. 1977.

Vol. 579: Combinatoire et Représentation du Groupe Symétrique, Strasbourg 1976. Proceedings 1976. Edité par D. Foata. IV, 339 pages. 1977.